Behaviormetrics: Quantitative Approaches to Human Behavior

Volume 7

Series Editor

Akinori Okada, Professor Emeritus, Rikkyo University, Tokyo, Japan

This series covers in their entirety the elements of behaviormetrics, a term that encompasses all quantitative approaches of research to disclose and understand human behavior in the broadest sense. The term includes the concept, theory, model, algorithm, method, and application of quantitative approaches from theoretical or conceptual studies to empirical or practical application studies to comprehend human behavior. The Behaviormetrics series deals with a wide range of topics of data analysis and of developing new models, algorithms, and methods to analyze these data.

The characteristics featured in the series have four aspects. The first is the variety of the methods utilized in data analysis and a newly developed method that includes not only standard or general statistical methods or psychometric methods traditionally used in data analysis, but also includes cluster analysis, multidimensional scaling, machine learning, corresponding analysis, biplot, network analysis and graph theory, conjoint measurement, biclustering, visualization, and data and web mining. The second aspect is the variety of types of data including ranking, categorical, preference, functional, angle, contextual, nominal, multi-mode multi-way, contextual, continuous, discrete, high-dimensional, and sparse data. The third comprises the varied procedures by which the data are collected: by survey, experiment, sensor devices, and purchase records, and other means. The fourth aspect of the Behaviormetrics series is the diversity of fields from which the data are derived, including marketing and consumer behavior, sociology, psychology, education, archaeology, medicine, economics, political and policy science, cognitive science, public administration, pharmacy, engineering, urban planning, agriculture and forestry science, and brain science.

In essence, the purpose of this series is to describe the new horizons opening up in behaviormetrics — approaches to understanding and disclosing human behaviors both in the analyses of diverse data by a wide range of methods and in the development of new methods to analyze these data.

More information about this series at http://www.springer.com/series/16001

Giuseppe Bove · Akinori Okada · Donatella Vicari

Methods for the Analysis of Asymmetric Proximity Data

Springer

Giuseppe Bove
Department of Education
Roma Tre University
Rome, Italy

Akinori Okada
Rikkyo University
Tokyo, Japan

Donatella Vicari
Department of Statistical Sciences
Sapienza University of Rome
Rome, Italy

ISSN 2524-4027 ISSN 2524-4035 (electronic)
Behaviormetrics: Quantitative Approaches to Human Behavior
ISBN 978-981-16-3171-9 ISBN 978-981-16-3172-6 (eBook)
https://doi.org/10.1007/978-981-16-3172-6

This Springer imprint is published by the registered company Springer Nature Singapore Pte Ltd.
The registered company address is: 152 Beach Road, #21-01/04 Gateway East, Singapore 189721, Singapore

To our families

Preface

This monograph originates from the long scientific collaboration between the three Authors on the subject of asymmetry in multidimensional scaling and cluster analysis, through many international meetings and workshops. An initiative of this collaboration was the Special Issue of the journal *Advances in Data Analysis and Classification* entitled *Analysis of Asymmetric Relationships*, published in the year 2018, now followed by this monograph.

The idea is not to write a second book on asymmetric structures of data analysis after the comprehensive exposition of Saito and Yadohisa (2005), but rather an agile monograph on the subject oriented to application in behavioural sciences. Many researchers in different fields collect proximity data such as brand switching or import–export data, flows and migration data and sociomatrices and need methods to explore the structure of their data. Therefore, the monograph is primarily intended as an updated overview of methods available for studying and analysing asymmetric relationships with a special focus on the practical relevance of the methods presented. Methods and results are always accompanied by illustrative graphical representations through real-life examples to be understandable to non-mathematicians and practitioners. The focus is on conceptual understanding and practical know-how rather than mathematical formalization and proofs. Specialized software is provided or referenced to in the final sections of some chapters, to help the computation of some of the models presented. The monograph can be also useful for graduate students interested in asymmetry in different fields of applications. Guidelines for reading are provided in Sect. 1.4.

We are indebted to many people we have shared time with to discuss asymmetry when preparing joint research papers or attending conferences in the last decades. Particularly useful and stimulating were the joint meetings and sessions organized between the Japanese and the Italian Groups of Classification and Data Analysis in 2012.

Special thanks go to Dr. Alessio Serafini for supporting the preparation of some R scripts and for discussing the software sections, to Prof. Tadashi Imaizumi for his help in preparing Sect. 5.5, and to Prof. Mark de Rooij who, with his careful reading, provided detailed and valuable suggestions that made the volume more understandable and accurate. The second author thanks Yosuko Hase for her help in English in Chap. 5.

Rome, Italy Giuseppe Bove
Tokyo, Japan Akinori Okada
Rome, Italy Donatella Vicari

Contents

Chapter 1
Introduction

1.1 Historical Background

In many disciplines such as psychology, sociology, marketing research, behavioural sciences and so on, systems of relationships between pairs of entities in a set are frequently studied. Examples are confusion matrices, sociomatrices, brand switching or import–export data, flows and migration data. Though many of these relationships are asymmetric in practice, asymmetry had been ignored for long time by methods of data analysis, with a few exceptions (e.g. Shepard [1]). A brief history of methods and procedures of multidimensional scaling (MDS) to deal with asymmetry in pairwise relationships data is provided in Saito and Yadohisa [2], Sect. 4.1.2. Here, only a few historical notes are recalled because of their relevance with respect to the methods of MDS and cluster analysis presented in the following chapters.

The problem to deal with asymmetry in MDS is already recognized in Kruskal [3], who proposed two strategies where asymmetry is removed when it is assumed to be due to random error.

The first strategy consists in averaging the corresponding entries in the observed square data matrix (*proximity matrix*) and then applying a distance model to the lower triangle of the symmetrized matrix; the second strategy directly approximates all the raw asymmetric proximities by a distance model.

Another approach frequently considered is to assume asymmetry resulting from conditionality of the data (e.g. row conditionality, where only values within rows are assumed to be comparable), allowing for different row transformations of the data, but still employing a symmetric model. In some applications, asymmetry might be ignored by fitting a symmetric model to the lower triangular part and/or the upper triangular part of the data matrix.

It is only from the mid-seventies that models and methods for the treatment of asymmetry as a phenomenon of interest start to be proposed. The increased awareness of the importance of asymmetry in proximity data comes mainly from psychologists and psychometricians. The development of new proposals to deal with asymmetry

G. Bove et al., *Methods for the Analysis of Asymmetric Proximity Data*,
Behaviormetrics: Quantitative Approaches to Human Behavior 7,
https://doi.org/10.1007/978-981-16-3172-6_1

is so intensive in those years that methods of asymmetric multidimensional scaling are already reviewed in Carroll and Arabie [4, pp. 619–621, 634–635].

One of the first proposals in the psychometric context comes from Young [5], who incorporates asymmetry in the Euclidean distance model allowing for differential weights for dimensions, depending on the rows of the proximity matrix (see Chap. 2 for a brief description of the model, and Young [6] for further details). Another asymmetric modification of the Euclidean distance model dated 1973 is the slide vector model, attributed to Kruskal by De Leeuw and Heiser [7]. The asymmetry is taken into account by adding a (slide) vector to the dimensionwise differences (see Chap. 2).

Tversky [8] and Tversky and Gati [9] reported many examples of psychological processes giving rise to asymmetric data, and assuming asymmetry as one of the reasons to advocate the superiority of non-spatial over spatial models. However, Krumhansl [10] eloquently replied to Tversky, proposing a *distance–density* model. Proximity is modelled as a function of the interpoint distance and of the local density of points in the surrounding region of the space, showing in this way that spatial models may still be relevant for asymmetric proximities. Holman [11] proposed a general class of models, known as the *similarity and bias* model (see also Nofosky [12]), where the proximity is expressed as a monotone function of a sum of a similarity function with two bias functions corresponding to rows and columns of the proximity matrix, respectively (see Chap. 3). A particular metric form of the similarity and bias model was proposed by Weeks and Bentler [13], where the similarity is a linear function of the Euclidean distance and the bias component is given by the difference between two scalar parameters. This model has been reconsidered in many studies on asymmetry until very recently.

The idea underlying many proposals of those years is that the proximity originates from symmetry, contaminated by response bias or other effects, that appropriate procedures can remove to enable the application of a symmetric model (e.g. a Euclidean distance model). This interpretation is quite evident, for instance, in Levin and Brown [15], who propose least-squares procedures for the symmetrization of a conditional proximity matrix, scaling rows and columns by multiplicative constants. As remarked in de Leeuw and Heiser [7], these approaches are closely related to the contingency table literature on "quasi-symmetry" and on social mobility tables (e.g. Bishop et al. [16], see also Chap. 3 in this monograph).

A second line of research can be identified in the context of factor analysis. An inner product-like model named DEDICOM which can handle one or more proximity matrices was proposed by Harshman [17]. A common set of dimensions for the rows and the columns of the proximity matrix is assumed, and asymmetries are accounted for by a set of indices of "directional relationships" which measure how each dimension affects any other dimension.

Chino [18] proposed another generalization of the inner product model, where the asymmetric relationships are split into the sum of a symmetric and a skew-symmetric component (a concept discussed in detail in Chap. 3). The method handles the two components simultaneously and provides a configuration for a joint representation. The paper of Chino published in *Behaviormetrika* has promoted a long tradition of papers on asymmetry in the Japanese school of Data Analysis (just a few names of contributors are T. Imaizumi, A. Okada, T. Saito, Y. Sato, H. Yadohisa) that extends to the present days (some contributions are described later).

Researchers in multivariate data analysis have also strongly contributed to the development of methods for asymmetric proximities. A key role has been played by John Gower, working at Rothamsted Experimental Station (Harpenden, UK). In two pioneering papers (Gower [19] and Constantine and Gower [20]), he presented new tools to represent square asymmetric matrices. In the first part of Gower [19], four (groups of) methods for visual inspection of asymmetric matrices based on eigen (singular)-value decomposition and least-squares approximation are provided. These methods are now included in the textbook of Cox and Cox [21], Sect. 4.8. In particular, it is pointed out that any method proposed for graphical representation of a rectangular data matrix could potentially be used to approximate an asymmetric proximity matrix, by considering rows and columns as two different sets of entities. A point of view shared also with other authors, (e.g. Carroll and Arabie [4, p. 619] and Kruskal and Carmone [22, p. 38]). So, for instance, bilinear models and unfolding models become useful tools for description of pairwise relationships (see Chap. 2 for examples of applications). In the second part of the work by Gower [19] relevant results concern the study of rank approximation and singular value decomposition (SVD) of skew-symmetric matrices, and the discovery of a new geometrical interpretation of the algebra of the SVD in terms of areas of triangles (see Chap. 3). In Gower [23], he himself provides a short account of how he became interested in analysing asymmetry, when he spent the year 1974 in Adelaide with the Commonwealth Scientific and Industrial Research Organisation (CSIRO), collaborating with Graham Constantine. In Constantine and Gower [20] this line of research is further developed, it is remarked that the additive decomposition of a proximity matrix in the symmetric and skew-symmetric components gives an orthogonal breakdown of the sum of squares, which suggests a separate analysis and modelling of the two components (the so-called *Constantine–Gower approach*, see Chap. 3). In the Constantine–Gower approach, the asymmetry is not considered as a random error to be removed, but it is an aspect of the data that can contain information worth to be described separately from symmetry.

Departures from symmetry are of special interest in the analysis of flow data as, for instance, migration data. Spatial migration tables have been frequently decomposed into a symmetric and a skew-symmetric part (e.g. Tobler [24]). In this decomposition, the skew-symmetry indicates the oriented net direction of the flow. Asymmetric models trying to identify the social and economic factors that render some areas more attractive than others were the central focus of several geographical studies (e.g. Tobler [25] and Constantine and Gower [26]).

A method for the simultaneous representation of the symmetric and the skew-symmetric components of an asymmetric square matrix was given in Escoufier and Grorud [27]. The method is based on the eigenvalue decomposition of a partitioned matrix whose blocks are based on the symmetric and the skew-symmetric components of the data matrix. Escoufier & Grorud method has close relationships with the methods proposed in Chino [18] and Gower [19] (as pointed out, for instance, in Chino and Shiraiwa [28]).

Less attention has been paid to clustering methodologies for asymmetric proximity data than the plenty of methodologies for representing objects in low-dimensional spaces, and most of the first proposals are placed within the framework of the hierarchical agglomerative methods.

Let us recall that the result obtained by a *hierarchical* method is a hierarchically-nested set of partitions of objects into groups (clusters) where objects in the same group are more similar to each other than objects in the other groups. In the agglomerative (also known as bottom-up) approach, each object forms a single cluster at the beginning and then the algorithm starts to merge the pair of most similar clusters until, step by step, all the objects belong to only one cluster. This allows to build a tree structure based on the data similarities which can be graphically displayed by a tree diagram denoted *dendrogram* (see Chap. 4 for some examples). Hierarchical methods differ in how they define the (dis)similarity between clusters in the merging step so that they belong to the same family of methods and Lance and Williams [29, 30] provided a general formula that allows to get most of the standard agglomerative algorithms by simply setting different parameters.

In the general framework of the agglomerative hierarchical clustering, the first proposals for clustering asymmetric data date back to the 1970s after the pioneering work by Hubert [31] who proposed a method to analyse asymmetric proximities that generalizes the single and complete linkage clustering [32]. Fujiwara [33] extended the method of Hubert by a more general algorithm which in turn was extended by Okada and Iwamoto [34] and where (a) a generalized dendrogram is proposed to visualize the nested clusters and (b) an algorithm is presented allowing to deal with the diagonal elements of the proximity matrix (self-proximities or proximity of an object to itself). Further extensions are due to Okada and Iwamoto [35], who proposed the weighted average algorithm for asymmetric clustering, and later to Yadohisa [36] who unified in a single general framework the asymmetric hierarchical algorithms by providing an updating formula for computing dissimilarities between clustered objects which incorporates asymmetry into its formulation. Such a formula, further extended by Takeuchi et al. [37], is actually an extension of the formula by Lance and Williams, so that it includes many algorithms specifically designed for asymmetric data and reduces to the standard agglomerative algorithms (i.e. single, complete, average linkage, …) in the case of symmetric proximities.

Following a different perspective, Ozawa [38] proposed a hierarchical clustering algorithm, called CLASSIC, by using nearest neighbour relations (NNR) and an iterative procedure to induce a hierarchy of combining clusters based on the nested sequence of NNRs.

Brossier [39] presented a clustering method which relies on the decomposition of the asymmetric proximity matrix into symmetric and skew-symmetric components. Due to the independence of the two components in a least-squares approach, Brossier proposed (a) to cluster objects from the symmetric part by using the standard average linkage algorithm and, separately, (b) to approximate the skew-symmetric part using structures for representing the imbalances (see Sect. 4.2.2 for details and examples).

It can be observed that only the average amounts of the proximities are actually involved in the clustering process, while the imbalances do not contribute to form the clusters, which opens the way to further developments towards joint clustering models that simultaneously take into account the two components.

It is worth to be noted that all the methods above consider the observed asymmetric proximity matrix according to its original form of square matrix where rows and columns refer to the same entities, while other proposals regard methods originally designed for standard rectangular matrices (i.e. objects $\times$ variables data matrices), and then adapted to asymmetric data. In some cases, the drawback is often the difficulty in handling a large number of objects which doubles because it is assumed that objects in rows and columns are different and even the graphical representations may become cumbersome and unclear.

With regard to *non-hierarchical* clustering methods, let us first recall that such methods provide one single grouping (i.e. partition, covering) of the objects in clusters by optimizing some criteria which generally aim at desirable properties such as internal cohesion of the groups and/or separation between groups. In such a framework, fewer contributions have been specifically designed for asymmetric data. Some proposals of non-hierarchical clustering methods for asymmetric data are actually extensions of the standard k-means algorithm [40–42] where the asymmetry is dealt with by defining appropriate either asymmetric proximities between probability measures or coefficients of asymmetry or an asymmetric Alpha–Beta divergence.

Also Okada and Yokoyama [43] proposed a sort of k-means-type algorithm (ACLUSKEW) based on the concept of dominant object [44] which is fully detailed and applied to real data in Sect. 4.3.4.

Following the line drawn by Brossier [39], Vicari [45] introduced a non-hierarchical clustering model which jointly handles symmetry and skew-symmetry because it relies on the decomposition of the asymmetric proximities into symmetric and skew-symmetric effects both decomposed into within and between cluster effects. In the resulting partition, objects in the same cluster share the same behaviours in terms of exchanges directed towards the other clusters, i.e. some clusters are mainly origins towards (destinations from) some other clusters. Moreover, in Vicari [46] an extension of the model (CLUSKEXT) is presented which incorporates possible external information to be used as covariates to explain the imbalances between objects. Both models are fully detailed and applied to real data in Sects. 4.3.1 and 4.3.2, respectively.

Moreover, in Vicari [47] the symmetric and skew-symmetric components are jointly fitted by two partitions nested one into the other in the sense that in the partition obtained from the skew-symmetries some objects are allowed to remain unassigned to any cluster: this indicates that such objects present a low degree of asymmetry (see Sect. 4.3.3).

Clustering methods for asymmetric proximity data originally proposed for standard (objects $\times$ variables) data are not explicitly presented in this monograph (see Chap. 5 in [2] for a review of other classes of methods).

All the lines of research presented in this section, developed mainly from the mid-seventies to the beginning of the eighties, were further developed until recently, and others were added later, as will be described in the following chapters of this monograph.

1.2 Types of Asymmetric Pairwise Relationships, Proximity Definition and Related Basic Concepts

Terms like relationship, association, correlation and similar are usually considered in statistics for the analysis of pairs of variables in a multivariate data matrix (e.g. subjects $\times$ variables). When more than two variables are involved, the goal of the analysis is to discover the structure underlying the relational system of all pairwise relationships among the variables (e.g. by using principal component or factor analysis methods).

However, in many different fields, interesting relational systems concern not only variables but also subjects, objects, stimuli, concepts and other entities. So, the word *proximity* is coined as a superordinate term that includes different concepts like nearness, relatedness, dependence, association, complementarity, substitutability and so on, defined for a set of subjects or variables or objects or stimuli, etc. A proximity can measure similarity or dissimilarity, with the obvious interpretation of measuring how similar or different a pair of subjects, variables, objects, etc. are.

There are a number of different ways of collecting proximity data [48, 49], which, according to Takane et al. [50], can be classified into two groups: *direct* judgements (e.g. ratings, rank order, sorting) or *indirect* methods (e.g. confusion data, co-occurrences, social interactions, flow data, profile dissimilarities, association measures). As an example of direct judgement, we can consider a rating task where in a marketing study a sample of subjects are asked to rate the degree of diversity of types of cars on a 10-point scale (i.e. $1 =$ most similar, $10 =$ most dissimilar). The dimensions (or attributes) of the comparison between cars can be (at least partially) explicitly specified to subjects before judgement, or can be hidden to be discovered by the analysis of the ratings.

As an example of indirect method, a reaction time experiment can be considered where two stimuli (sounds, words or symbols) are presented at a time, and subjects evaluate whether the stimuli are "the same" or "different" as quickly as possible. Registered average reaction times can be considered measures of confusion (similarity type) between stimuli. Other measures of confusion between stimuli can be obtained as percentages in the so-called stimulus–stimulus or stimulus–response experiments (e.g. Kruskal and Wish [48]). Stimuli can be also rated on a number of given attributes collected in a multivariate data matrix (stimuli $\times$ attributes), and proximities can be defined in many different ways as dissimilarities (e.g. Euclidean distances) or similarities (e.g. association measures) between pairs of stimuli (the reader is referred to Cox and Cox [21] and Everitt and Rabe-Hesketh [51] for a list and a discussion of possible choices of (dis)similarity measures).

The observed proximity judgements can be *asymmetric*. As already remarked, Tversky [8] and Tversky and Gati [9] questioned the assumptions of the geometric representation of proximity data, including the symmetric property, and provided examples of contexts where pairwise (dis)similarity ratings can be asymmetric (e.g. in evaluating the similarities between Nations, the judged similarity of North Korea to China might outweigh the judged similarity of China to North Korea). Sociomatrices of ratings representing relations as liking/disliking, friends/enemies within a group of people are other examples of asymmetric proximities, because person i could express a higher like or dislike rating towards person j than the reverse. An example of sociomatrix is presented in the next section.

Asymmetry affects many situations where proximities are obtained by indirect methods. In stimulus identification experiments, subjects are asked to evaluate whether two stimuli presented in a given order were the same, and the percentage of times stimulus i is "confused" with stimulus j is generally different from the percentage of times stimulus j is "confused" with stimulus i. A small example of proximities obtained in a stimulus identification experiment will be studied in Chap. 3. Data concerning information/communication flows, travel flows or other types of transactions can give rise to asymmetric proximities. For instance, pairwise relationships between members in organizational structures (e.g. employees in a firm, students at school, friends in a social network) are frequently asymmetric (e.g. "A sends an e-mail to B"). An example concerning the relationship of the type "A approaches B for help and advice" among managers is provided in the next section. Trade or migration flows among countries, import/export flows and brand switching among customers show usually large imbalance. A small example of import/export flows between countries will be studied in Chap. 2.

A broad definition of proximity

All types of proximities presented in this and in the following sections meet the following definition: *a proximity is a measure of relationship between ordered pairs of elements in a given set*. From now on, a proximity is generally denoted by the symbol ω, and regardless of its level of measurement it is assumed as a real function defined on ordered pairs of elements of a set O. The n elements of O are labelled by the integers 1, 2, …, n, so that the values of ω on the pairs (i, j) are indicated ω_{ij} $(i, j = 1, 2, \ldots, n)$. All values of ω can be collected in a $(n \times n)$ data matrix (i.e. a square array of numbers organized in n rows and n columns) indicated by $\mathbf{\Omega} = \left(\omega_{ij}\right)$. As for other types of data matrices, some values ω_{ij} can be missing observations or not defined (e.g. very frequently diagonal entries ω_{ii} are not defined or not of interest).

According to the classification by Carroll and Arabie [4], $\mathbf{\Omega}$ is a one-mode two-way matrix, because it concerns one set of elements (one-mode) and entries ω_{ij} have two indices i, j (two-way). When several proximity matrices are observed on several occasions or situations (e.g. several times or places), each matrix will be indicated by $\mathbf{\Omega}_k = \left(\omega_{ijk}\right)$, where the subscript k denotes the occasion of observation $(k = 1, 2, \ldots, K)$. In a geometric perspective, the parallelepiped formed by the K

square matrices $\mathbf{\Omega}_k$ depicts a two-mode (the second set represents the occasions), three-way (entries ω_{ijk} have three indices) matrix.

A proximity measure may be *bipolar* (when the measure has three designated markers on its scale: a minimal value, a neutral value and a maximal value) or *unipolar* (nonnegative). The Pearson (product-moment) correlation coefficient, as other association measures, is an example of bipolar proximity (markers are -1, 0 and 1), while a confusion percentage is an example of unipolar proximity.

A proximity measure may be asymmetric (at least for a pair (i, j) $\omega_{ij} \neq \omega_{ji}$ holds) or symmetric ($\omega_{ij} = \omega_{ji}$, $(i, j = 1, 2, \ldots, n)$). In some parts of this monograph, when required by the specific problem treated, an asymmetric proximity will be denoted by either ω^{sim} or ω^{dis} when measuring similarity or dissimilarity, respectively.

A *dissimilarity* is a unipolar proximity where the dissimilarity of an element with itself is required to be zero. The $(n \times n)$ matrix collecting the values of a symmetric dissimilarity will be denoted $\Delta = (\delta_{ij})$. A *similarity* is a unipolar proximity where the similarity of an element with itself is required to be greater than zero and usually a maximum (in this case the similarity can be scaled so that the maximum similarity is equal to one, with the similarity of an element with itself set equal to one). The $(n \times n)$ matrix collecting the values of a symmetric similarity will be indicated by $\mathbf{S} = (s_{ij})$.

From now on, the terms element, entity and object will be used as synonyms.

Data collection and derived proximity matrices

Data observed by direct judgments or indirect methods can give rise to proximity matrices in different ways, also depending on the aim of the study. For instance, in sorting tasks a subject is required to sort a set of stimuli into as many categories as he wants according to their similarities, and the stimuli within the same category are considered more similar to each other than stimuli assigned to different categories. As a result, a binary proximity matrix (with entries 0 and 1) can be built, where entries equal to ones correspond to pairs of stimuli belonging to the same categories. Proximity matrices provided by several subjects can be analysed separately by cluster methods to compare the different classifications. Furthermore, the two-mode three-way proximity matrix formed by all the individual matrices can be analysed by multidimensional scaling methods, to provide graphical representations and to discover "psychological" dimensions (if any) the subjects implicitly assume to classify the stimuli. Moreover, the subject matrices can be aggregated into a single one-mode two-way proximity matrix containing the number of subjects assigning the pairs of stimuli to the same category, and the aggregated matrix can be analysed by a graphical or a cluster method.

Analogously, (dis)similarities between profiles can be defined in many different ways (e.g. Cox and Cox [21]) for multivariate data (e.g. attributes for a set of stimuli, features for market goods), and clustering methods can be applied to classify stimuli or other entities in several "typologies". The nature of the data and the aim of the analysis strongly influence the choice of similarity or dissimilarity coefficients (e.g. Gower and Legendre [52]).

Structures for proximity data

Proximities can be analysed by different types of models and methods. The choice of a specific model depends on the purpose of the analysis, on the data collection process, and, more generally, on the amount of information available on the phenomenon under investigation. In some situations, the theory suggests an explicit substantive model for the data; examples can be found in both experimental and non-experimental studies [53]. For instance, in perception studies, the knowledge of the attributes that determine the stimulus proximities can suggest peculiar spatial models representing psychological composition rules (e.g. Borg and Groenen [49], Sect. 1.4).

When no substantive model is readily available and there is no a priori reason to prefer any particular model, multidimensional scaling and cluster analysis methods can be considered in an exploratory perspective for data visualization, dimensionality reduction, scale development, classification. Specifically, graphical displays are useful for a preliminary investigation of the data matrix that analogously to scatterplots "allows the data to speak for themselves".

Hartigan [54] provides 12 possible *proximity structures* that might need to be considered to describe an observed symmetric proximity (similarity or dissimilarity). A subset of such structures will be considered in the following chapters, a list in order of decreasing complexity follows where the symbol a is generally used for representing the structure defined on the ordered pairs of elements of the set O.

(a) a is a Euclidean distance
(b) a is a distance
(c) a is symmetric real valued
(d) a is real valued
(e) a is a complete "proximity" order
(f) a is a tree
(g) a is a partition of the set O into sets of similar elements.

The observed proximities discussed in the following chapters usually satisfy one of the conditions (c), (d), (e) and rarely meet conditions (a) and (b). In particular, condition (e) means that the measurement level of the observed proximities is ordinal: given two pairs of elements we can always say which is more (dis)similar or whether their (dis)similarities are the same. Structures (f) and (g) are non-metric and will be considered in detail in Chap. 4.

A matrix collecting distances denoted by $\mathbf{D} = \left(d_{ij}\right)$ has zeros as diagonal entries, it is symmetric $d_{ij} = d_{ji}$, $(i, j = 1, 2, \ldots, n)$, and, for any i and j, $d_{ij} \leq d_{ik} + d_{jk}$ for all k (the *triangle inequality*). A distance is Euclidean when it represents the spacing of a set of n points in a Euclidean space (examples are provided in Chap. 2).

A special distance is the *ultrametric* distance which satisfies a stronger conditions on the triplets such that, for i and j, $d_{ij} \leq \max\left\{d_{ik}, d_{jk}\right\}$ for all k (*ultrametric inequality*).

Models for representing asymmetric proximities will be based on appropriate modifications or combinations of the structures considered. The values of a model defined on the pairs of elements (i, j) of the set O will be indicated by a_{ij}, and collected in matrix $\mathbf{A} = (a_{ij})$. For instance, for one of the models discussed in Chap. 3 the structure a is a combination of types (a) and (d), and the entries a_{ij} are the sum of a Euclidean distance and a value computed as the difference between two scalar parameters (an example of a *hybrid model* according to Gower [53]). As it will be shown, the two structures of the hybrid model provide a joint graphical representation in a plane that allows an easy detection of symmetry and asymmetry of the proximity data. Analogously, hybrid models for symmetric proximities [55] can combine spatial structures with tree structures, gaining a deeper insight into the data matrix after superimposing the tree information in the spatial configuration (e.g. Shepard [56]).

According to the motivation and aims of data analysis, models and methods based on spatial, tree, cluster or hybrid structures will be considered in the following chapters to approximate and represent a proximity matrix. In particular, spatial structures will be presented mainly in Chaps. 2, 3 and 5, while tree and cluster structures will be discussed in Chap. 4.

Following a descriptive approach, the choice of the model (or the combined structure) to represent a proximity is performed by the *least squares* (*LS*) criterion that can be interpreted as a distance between the observed matrix $\mathbf{\Omega}$ (or its *admissible transformations* determined by its measurement level) and matrix $\mathbf{A}$ representing the class of the candidate models (examples of *LS* criteria are provided in Sect. 2.3.1). The model chosen in the class is the one minimizing the *LS* criterion. Except in some very special cases, the models do not perfectly represent the proximities, so that the *LS* criterion is taken as a measure of the approximation obtained by the model chosen.

Proximities may imply a “dominance” pairwise relationship (e.g. import/export data, sociomatrices) and can present intransitivities when for one or more triples (i, j, k) $\omega_{ij} > \omega_{ji}$, $\omega_{jk} > \omega_{kj}$ and $\omega_{ik} < \omega_{ki}$ hold. These intransitivities are also called *circular triads* because when the triple (i, j, k) is represented in a graph, the directed arrows indicating the signs of the relations form a circle. Some models are not able to represent circular triads (e.g. all models assuming linear forms of asymmetry), so it can be useful to measure in advance the amount of circular triads in the observed proximity matrix by an appropriate coefficient [2, 57, 58].

Transformations and decompositions of proximity data

Proximities defined by direct judgments or indirect methods can be transformed for several reasons.

Proximities which measure a similarity relationship ω_{ij}^{sim} need to be converted into dissimilarities ω_{ij}^{dis} before fitting models based on distances by using simple transformations such as

$$\omega_{ij}^{\text{dis}} = c - \omega_{ij}^{\text{sim}}, \tag{1.1}$$

for some constant c. For instance, an observed proportion of confusion between stimuli (an example of similarity relationship) is simply converted into a proximity measuring a dissimilarity relationship by the transformation

$$\omega_{ij}^{\text{dis}} = 1 - \omega_{ij}^{\text{sim}}. \tag{1.2}$$

In the case of percentages of confusion, it can be $c = 100$. Another example concerns the Pearson correlation coefficient (and other analogous normalized association measures) that may be converted into a dissimilarity measure by the transformation

$$\omega_{ij}^{\text{dis}} = \sqrt{c\left(1 - \omega_{ij}^{\text{sim}}\right)}, \tag{1.3}$$

for some constant c (e.g. $c = 2$).

Transformations can be also derived by a substantive model. For instance, in some applications (e.g. Zielman and Heiser [59] and Groenen and Heiser [60]) the following transformation has been considered

$$\omega_{ij}^{\text{dis}} = \sqrt{\frac{m_i m_j}{\omega_{ij}^{\text{sim}}}}, \tag{1.4}$$

where m_i and m_j are defined according to the application field (e.g. they may correspond to the row and column sums in a brand switching matrix where ω_{ij}^{sim} is the raw association frequency). The transformation stems from the so-called *gravity model* [61], a substantive model derived from gravitational studies in physics. For other examples of transformations of similarities into dissimilarities, the reader is referred to Mair et al. [62], Table 2.

Conversely, a transformation from proximities measuring dissimilarities to proximities measuring similarities arises from geometry (*cosine law*), and it is frequently applied in computational programs performing multidimensional scaling. The transformation has its origin in a work of Young and Householder [63], who showed how distances between points in a Euclidean space can be preserved by using an appropriate system of point coordinates. The transformation is

$$\omega_{ij}^{\text{sim}} = c_{ij} - c_{i.} - c_{.j} + c_{..} \tag{1.5}$$

where $c_{ij} = -\frac{1}{2}\left(\omega_{ij}^{\text{dis}}\right)^2$, and $c_{i.}$, $c_{.j}$, $c_{..}$ denote the row, column and total averages of matrix $\mathrm{C} = (c_{ij})$, respectively. It can be shown that when ω_{ij}^{dis} are Euclidean distances, then ω_{ij}^{sim} are the corresponding inner products (see Chap. 2 for the definition of inner product). Torgerson [64] used (1.5) to propose a technique called *classical* multidimensional scaling, whose details can be found, for instance, in Everitt and Rabe-Hesketh [51].

Proximities represented by frequencies (e.g. switching data, flow data) can be transformed to reduce large differences among the entries due to disturbing factors. For instance, in car switching data the large differences in level sums (row-plus-column sums) are generally due to extraneous factors, such as market share [65]. Transformations where the frequencies are first transformed in row proportions and then divided by the product of the corresponding row and column arithmetic means have been considered for brand switching data (e.g. DeSarbo [66]). Alternatively, iterative computational procedures can adjust rows and columns of the data matrix until the sum of all entries of the market segments are equal (i.e. the sum of row i plus the sum of column i equals the sum of row j plus the sum of column j, for all i, j), for applications the reader is referred to Harshman et al. [65] and Rocci and Bove [67].

Transformations can be also suggested (or performed) by appropriate statistical models. For instance, when proximities are frequencies, Heiser and Busing [68] considered the transformation

$$\omega_{ij}^{*} = \frac{\omega_{ij}}{\sqrt{\omega_{ii}\omega_{jj}}} \tag{1.6}$$

where ω_{ij}^{*} are the transformed proximities. A property of this sort of standardization is that if the simple model $\omega_{ij} = \alpha_i \alpha_j \theta_{ij}$ holds, with α_i main effect parameters (row/column effects) and θ_{ij} interaction parameters between rows and columns with $\theta_{ii} = 1$, then $\omega_{ij}^{*} = \theta_{ij}$ holds. It follows that a graphical representation or a classification based on the transformed proximities ω_{ij}^{*} is actually a way to describe the parameters of a statistical model. Other authors followed similar approaches, firstly applying a statistical model to a proximity matrix (in this perspective, a model may be viewed as a transformation of the proximities into the set of parameters of the model) and then representing the parameters (and possibly the residuals) by multidimensional scaling and bilinear models (see Caussinus and de Falguerolles [69], for an example of application of this approach to the Thomas sociomatrix to be presented in Sect. 1.3).

Transformations consisting in adding an appropriate constant to the extra-diagonal dissimilarities were also considered to render a dissimilarity into a (Euclidean) distance matrix to be represented in a low-dimensional space. Examples can be found for instance in Gower and Legendre [52] or Cox and Cox [21]. However, it has been observed that transformations by large additive constants usually lead to a large number of dimensions.

Proximities, even if expressed numerically, frequently may contain only order information. For instance, this is the case of proximities obtained from the data collection method called conditional rank orders [50], when subjects are asked to provide a (dis)similarity rank order of a set of stimuli with respect to a standard stimulus (one of the stimuli acts in turn as a standard stimulus). In such situations, the level of measurement of the proximity can be taken into consideration by incorporating a *monotonic transformation* (e.g. ordinal or linear) of the proximities in the computational algorithm for model estimation. When proximities are similarities, the transformation is required to be monotonically decreasing (or linear with negative slope for proximities measured at ratio or interval measurement levels). We will return on these aspects in Chap. 2.

Additive and multiplicative decompositions of proximities have been usefully studied and applied in various contexts. Strategies that split an asymmetric proximity matrix additively into two parts (i.e. a symmetric and a skew-symmetric one) to be either separately or jointly represented will be studied in detail in Chaps. 3 and 4. Multiplicative decompositions of a proximity into a symmetric and an antisymmetric factor were considered in studies concerning stimulus identification experiments and dominance relational data, to develop response processes and choice models (the reader is referred to Heiser and Busing [68] for these approaches).

1.3 Examples of Asymmetric Proximity Matrices

In this section, some examples of proximity matrices taken from different fields of application are provided. The data will be analysed in detail in the following chapters to show capabilities and limitations of the models and methods that will be presented.

Managers data [70]

Relationships between members in organizational structures (e.g. employees in a firm, students at school, friends on web) are frequently asymmetric (e.g. "A is a friend of B", "A sends an e-mail to B"). Krackhardt [70] reported an empirical study concerning relationships among 21 managers of a small manufacturing organization, producing high-tech machinery for other companies. The management consists of one president, four vice presidents and sixteen supervisors. Each vice president heads up a department, and each of the sixteen supervisors belongs to one of the four departments. For each manager, the position within the Department he belongs to is available as reported in Table 1.1 (reproduced by Okada [71], Table 1, p. 221). The kind of relationship to be analysed (of the type "A approaches B for help and advice") was submitted to all 21 managers, who were presented with a questionnaire of 21 questions.

Each question was contained in a separate page, and it asked, "Who would manager X go to for help or advice at work?". The rest of the page consisted of the list of the names of the remaining 20 managers, and the respondent was instructed

Table 1.1 Managers data: department composition

Department	Vice president	Supervisors
1	21	6, 8, 12, 17
2	14	3, 5, 9, 13, 15, 19, 20
3	18	10, 11
4	2	1, 4, 16

to put a check mark near the names of the managers to whom manager X was likely to go for advice.

Therefore, the original data consist of 21 square (21 × 21) matrices (termed slices) of dichotomous values 0/1 (available in Krackhardt [70], Appendix A, p. 129), each provided by one manager. The set of slices (an example of a one-mode three-way array) can be aggregated and analysed by different approaches (for details, the reader is referred to Krackhardt [70], Wasserman and Faust [72], Bove [73], Okada [71]). Frequently, the aggregated proximity matrix resulting by summing all the slices is considered. Table 1.2 provides the asymmetric proximity matrix obtained as the sum of the 21 slices, where each off-diagonal entry (i, j) represents the number of managers who responded that manager i goes to manager j for help or advice at work.

Diagonal entries are not defined. The proximity is assumed as a measure of closeness (or similarity) between managers. Graphical representations of the data are provided in Chaps. 2 and 4 in order to analyse the relationships between managers, and the centrality of the role of president and vice president.

Thomas Sociomatrix [74]

Sociomatrices are an interesting example of one-mode two-way proximity matrix, where positive or negative affective relations in a group of people as liking/disliking, friendship/hostility are collected. Sociomatrices are frequently asymmetric, because subject i might rate subject j with a higher like or dislike rating than vice versa. Table 1.3 is an example of sociomatrix concerning liking ratings (between 0 = min and 20 = max) that 24 students in a class assigned to each of their classmates. Diagonal entries are not defined. The proximity is regarded as a measure of asymmetric similarity between students. These data have been studied with different approaches by Cailliez and Pages [74, pp. 294–302]) and Caussinus and de Falguerolles [69], to identify the pattern of affinities between the 24 classmates. The results of the application of some methods of asymmetric multidimensional scaling to this data will be presented in Chap. 3.

European Languages data [75]

Data come from the Special Eurobarometer-243 survey [75] on Europeans and multilingualism within the European Union. A sample of Europeans were asked: “Which languages do you speak well enough in order to be able to have a conversation, excluding your mother tongue?” and the percentage ω_{ij} of citizens from country i who responded to speak language j is reported in Table 1.4 for 16 countries and their corresponding languages.

In Table 1.4, countries and languages are referred to by their official abbreviations and by short labels, respectively, as follows:

Countries

Czech Republic (CZ), Denmark (DK), Germany (DE), Spain (ES), France (FR), Italy (IT), Hungary (HU), The Netherlands (NL), Poland (PL), Portugal (PT), Slovakia (SK), Finland (FI), Sweden (SE), The United Kingdom (UK), Bulgaria (BG), Romania (RO).

Table 1.2 Proximity matrix between 21 managers in a firm [70]

Manager	m1	m2	m3	m4	m5	m6	m7	m8	m9	m10	m11	m12	m13	m14	m15	m16	m17	m18	m19	m20	m21
1	–	21	7	11	5	1	1	7	2	6	13	0	5	2	2	11	2	19	5	0	2
2	10	–	7	11	5	8	21	9	2	4	6	7	4	14	2	7	6	15	5	0	15
3	8	16	–	2	2	3	12	2	3	8	11	2	0	20	1	2	3	14	2	2	6
4	9	18	0	–	0	7	2	15	0	6	5	6	0	0	0	5	2	11	0	1	11
5	7	14	4	1	–	3	12	2	2	4	12	2	4	20	2	4	6	16	10	10	8
6	2	11	1	7	2	–	15	4	2	1	2	11	1	7	2	2	9	1	2	2	21
7	2	17	6	2	3	8	–	3	3	2	11	3	2	19	3	2	6	13	3	6	20
8	7	16	4	11	0	6	10	–	1	7	10	5	0	2	0	4	5	14	0	2	18
9	5	11	7	0	11	5	8	2	–	3	5	3	6	21	6	3	5	12	11	6	6
10	3	8	7	3	4	1	8	8	1	–	8	0	4	3	2	9	2	20	4	6	0
11	9	15	6	3	6	1	20	7	1	3	–	0	5	11	4	1	1	19	5	2	1
12	1	8	0	10	0	15	8	7	0	0	0	–	0	4	1	0	7	3	0	1	20
13	9	11	4	0	17	2	7	1	11	4	11	0	–	21	5	4	2	16	15	8	1
14	2	16	9	1	10	5	20	4	11	4	9	4	7	–	9	1	4	17	10	10	17
15	4	14	2	1	8	4	10	2	6	2	7	1	3	21	–	4	3	17	11	12	5
16	13	19	2	11	0	3	0	8	0	11	3	0	0	3	0	–	2	15	0	3	1
17	6	15	5	4	5	15	17	6	2	1	3	12	5	9	3	2	–	6	5	3	21
18	9	17	9	5	5	5	19	8	3	11	15	3	5	16	5	8	5	–	5	8	10
19	8	14	9	0	13	2	9	2	4	7	12	0	4	21	7	3	5	16	–	11	2
20	6	15	6	5	9	8	11	5	5	8	10	5	6	21	9	8	4	16	8	–	6
21	0	16	4	4	0	11	21	6	0	0	0	10	0	15	0	0	10	6	0	3	–

Table 1.3 Thomas sociomatrix [74]

	s1	s2	s3	s4	s5	s6	s7	s8	s9	s10	s11	s12	s13	s14	s15	s16	s17	s18	s19	s20	s21	s22	s23	s24
s1	–	5	12	14	15	3	18	10	10	10	14	14	5	20	17	15	19	17	10	16	12	16	15	12
s2	10	–	17	16	14	20	17	19	13	17	11	13	8	10	6	1	17	17	16	17	13	17	17	19
s3	15	9	–	11	19	17	13	7	16	8	9	2	3	15	14	10	13	14	14	10	10	16	16	17
s4	15	12	11	–	10	17	12	16	13	15	12	10	8	8	11	11	13	13	17	16	14	14	15	18
s5	14	4	19	11	–	17	15	8	16	9	3	1	2	15	10	10	10	3	13	9	13	15	15	13
s6	9	19	18	13	18	–	10	5	12	2	2	1	5	2	12	3	17	15	15	11	11	12	16	14
s7	16	15	14	10	15	15	–	10	16	10	8	2	1	16	6	13	20	2	8	10	8	18	16	14
s8	17	18	15	13	16	18	12	–	15	16	17	13	11	16	10	14	16	15	12	19	18	18	17	18
s9	14	14	12	10	8	10	15	10	–	9	14	8	5	14	13	1	14	9	10	11	13	18	19	15
s10	19	10	15	12	10	15	13	20	15	–	12	8	12	17	12	3	12	12	10	20	14	15	15	15
s11	19	12	15	12	15	8	14	14	17	9	–	15	5	19	16	19	17	16	10	10	19	18	18	15
s12	20	15	14	17	16	13	13	10	17	13	14	–	13	19	10	12	11	19	8	14	14	18	18	9
s13	13	2	15	11	3	19	10	12	12	18	8	19	–	12	16	12	12	20	4	18	10	15	16	17
s14	20	5	13	14	15	5	18	16	10	10	14	13	5	–	17	16	17	13	10	16	11	15	15	13
s15	14	10	8	6	14	10	5	9	15	7	1	6	3	13	–	13	19	14	1	10	7	18	17	2
s16	10	10	5	3	6	6	8	7	9	6	15	12	1	11	10	–	8	10	4	5	11	13	8	3
s17	18	17	10	10	14	17	20	17	16	10	16	1	1	18	17	15	–	1	10	9	10	19	17	13
s18	15	12	10	10	3	16	10	13	10	10	11	20	17	15	18	10	10	–	10	13	19	14	14	10
s19	12	14	14	13	17	18	14	12	13	11	10	5	12	12	12	11	11	10	–	11	11	11	12	14
s20	18	9	11	16	6	15	13	20	17	19	12	8	13	18	19	10	11	11	14	–	14	19	17	15
s21	20	8	14	12	16	8	7	10	10	6	19	11	4	19	7	16	11	16	2	16	–	14	13	12

(continued)

Table 1.3 (continued)

	s1	s2	s3	s4	s5	s6	s7	s8	s9	s10	s11	s12	s13	s14	s15	s16	s17	s18	s19	s20	s21	s22	s23	s24
s22	14	13	13	12	13	10	16	12	18	12	10	12	12	14	18	9	15	10	9	13	10	–	19	18
s23	14	13	13	9	10	11	15	9	19	10	12	12	6	14	13	12	13	13	13	12	13	18	–	16
s24	12	15	16	19	15	15	11	12	13	13	4	12	13	12	12	7	12	11	14	13	11	17	17	–

Table 1.4 Percentage of people (%) in European countries speaking foreign languages (Special Eurobarometer-243, [75])

	Czech	Dan	Germ	Span	French	Ital	Hung	Dutch	Pol	Port	Slov	Finn	Swed	Engl	Bulg	Rom
CZ	–	0	19	0	1	0	0	0	1	0	9	0	0	19	0	0
DK	0	–	8	0	1	0	0	0	0	0	0	0	1	76	0	0
DE	0	0	–	1	1	1	0	1	1	0	0	0	0	49	0	0
ES	0	0	1	–	5	0	0	0	0	1	0	0	0	19	0	0
FR	0	0	4	5	–	2	0	0	0	0	0	0	0	31	0	0
IT	0	0	2	1	8	–	0	0	0	0	0	0	0	24	0	0
HU	0	0	20	0	0	0	–	0	0	0	1	0	0	13	0	2
NL	0	0	23	0	2	0	0	–	0	0	0	0	0	60	0	0
PL	1	0	11	0	1	1	0	0	–	0	0	0	0	23	0	0
PT	0	0	2	3	8	0	0	0	0	–	0	0	0	28	0	0
SK	14	0	14	0	0	0	5	0	0	0	–	0	0	20	0	0
FI	0	0	4	0	0	0	0	0	0	0	0	–	8	52	0	0
SE	0	2	2	0	0	0	0	0	0	0	0	0	–	83	0	0
UK	0	0	5	4	18	0	0	0	0	0	0	0	0	–	0	0
BG	0	0	5	0	3	0	0	0	0	0	0	0	0	16	–	1
RO	0	0	2	1	11	1	2	0	0	0	0	0	0	23	0	–

Table 1.5 Additional information on European countries (Special Eurobarometer-243, [75])

	CZ	DK	DE	ES	FR	IT	HU	NL	PL	PT	SK	FI	SE	UK	BG	RO
X_1	72	77	84	77	76	66	70	58	75	79	80	80	73	55	78	72
X_2	61	88	67	44	51	41	42	91	57	42	97	69	90	38	59	47
X_3	2	2	8	9	6	1	0	3	0	0	9	4	3	7	8	3
X_4	34	42	42	25	32	31	31	50	28	20	42	39	40	21	23	30

Languages

Czech (Czech), Danish (Dan), German (Germ), Spanish (Span), French (French), Italian (Ital), Hungarian (Hung), Dutch (Dutch), Poland (Pol), Portuguese (Port), Slovak (Slov), Finnish (Finn), Swedish (Swed), English (Engl), Bulgarian (Bulg), Romanian (Rom).

Diagonal entries in Table 1.4 are not of interest. Data are an example of asymmetric similarities between countries/languages. European languages data are analysed in Chap. 2 by a bilinear method and Chap. 4 by clustering methods. A submatrix related to only seven selected countries is used throughout Chap. 4 as an example to better illustrate the specific clustering methods.

Europeans were also asked several questions and the percentages of their responses are reported in Table 1.5 which contains additional information on the sixteen countries:

X_1—people who responded that it is important that young people learn other languages at school or university for improving job opportunities
X_2—people speaking at least one foreign language
X_3—people speaking the language in the country they live as a foreign language
X_4—people who responded that the main reason for learning a new language is to use at work (including travelling abroad on business).

The external information on countries is used in Chap. 4 to get a better insight on the multilingualism and the attitudes of Europeans towards foreign languages.

Intergenerational occupational mobility data [76]

Data on intergenerational mobility among eight occupational categories consist of occupational mobility from fathers to sons among eight categories in four years (1955, 1965, 1975 and 1985) in Japan (Seiyama et al. [76, pp. 46–47], Table 2.12). The eight occupational categories are

Professional occupations
Non-manual occupations employed by **large** enterprises
Non-manual occupations employed by **small** enterprises
Non-manual self-employed occupations
Manual occupations employed by **large** enterprises
Manual occupations employed by **small** enterprises
Manual self-employed occupations and

Farm occupations.

The words in bold are used to denote the occupational category henceforth. The original data form four (8 × 8) tables and are reported in Table 1.6. Each table of Table 1.6 contains the data from the surveys conducted in 1955, 1965, 1975 and 1985, respectively, and any cell of each table represents the number of intergenerational occupational movements from fathers in the row category to sons in the column category in each year, so that data represent asymmetric similarities between occupational categories.

An application of two-mode three-way asymmetric multidimensional scaling [77] to this data is presented in Chap. 5.

1.4 Aims and Overview of the Book

This monograph is devoted to the analysis of asymmetric proximities mainly oriented to practical applications. The aim is conceptual understanding and practical know-how rather than mathematical formalization and proofs. The book is planned as an agile monograph covering an up-to-date overview of methods to analyse asymmetric proximity data by building graphical representations which are illustrated through real life examples. For this reason, each of the following chapters contains a final section of applications and it usually starts with a small simple data set which allows to present the different models and methods throughout the chapter.

The monograph is intended not only for research workers and professionals operating in the most varied fields of application but also for graduate students who are interested in asymmetry. A basic knowledge of geometry and matrix algebra helps the reading of Chaps. 2 and 3, but the necessary concepts will still be recalled. Chapters 4 and 5 present more advanced arguments and need at least a basic knowledge of the standard cluster analysis and multidimensional scaling algorithms. Throughout the book, sections with the title marked with an asterisk contain more complex topics that are only briefly described with suggestions for further readings.

To begin with, in Chap. 2 after a brief recall to the main aspects of the graphical representation of a rectangular matrix, two classes of methods to represent proximity matrices are reviewed: bilinear and distance-like methods. The main steps for the bilinear model estimation are presented by studying the properties of the singular value decomposition of a proximity matrix. The principal concepts concerning the distance models for the analysis of symmetric proximity matrices are recalled, then some distance-like methods dealing with the direct representation of the asymmetric proximities are presented. The chapter ends with two applications to real data and a software section to present some programs available for models estimation.

In Chap. 3, the properties of the decomposition of a square data matrix in its symmetric and skew-symmetric components are firstly presented and discussed. The orthogonal breakdown of the sum of squares of the proximities characterizing the decomposition is emphasized. Afterwards, an overview of models and

Table 1.6 Original intergenerational occupational mobility

Father's occupation	Son's occupation year 1955							
	1	2	3	4	5	6	7	8
1 Professional	42	12	8	2	1	3	6	9
2 Non-manual large	14	19	8	13	5	6	5	17
3 Non-manual small	11	6	5	8	3	6	3	0
4 Non-manual self	21	30	20	72	11	13	26	14
5 Manual large	3	3	10	3	21	12	7	5
6 Manual small	2	5	3	6	14	22	13	11
7 Manual self	9	24	14	18	25	42	88	27
8 Farm	37	59	26	71	67	64	81	686
Father's occupation	Son's occupation year 1965							
	1	2	3	4	5	6	7	8
1 Professional	28	16	7	2	5	8	6	2
2 Non-manual large	22	44	15	17	14	10	9	10
3 Non-manual small	16	11	14	7	9	7	3	1
4 Non-manual self	15	41	21	94	14	20	17	9
5 Manual large	6	20	12	1	26	22	7	5
6 Manual small	1	9	13	5	18	32	12	11
7 Manual self	4	39	22	30	25	39	78	12
8 Farm	44	93	56	69	107	166	73	339
Father's occupation	Son's occupation year 1975							
	1	2	3	4	5	6	7	8
1 Professional	44	26	18	10	5	6	4	5
2 Non-manual large	22	74	26	29	15	20	10	9
3 Non-manual small	15	17	16	7	6	18	1	3
4 Non-manual self	22	58	44	103	24	32	38	8
5 Manual large	11	18	12	8	29	31	6	5
6 Manual small	7	15	19	7	23	56	10	5
7 Manual self	15	33	29	28	23	44	88	14
8 Farm	44	99	98	74	118	236	83	325
Father's occupation	Son's occupation year 1985							
	1	2	3	4	5	6	7	8
1 Professional	40	29	22	5	3	9	1	0
2 Non-manual large	31	84	33	20	19	20	8	3
3 Non-manual small	15	18	21	6	5	22	4	0
4 Non-manual self	27	38	39	99	17	26	18	4
5 Manual large	19	25	16	3	32	21	8	0

(continued)

Table 1.6 (continued)

Father's occupation	Son's occupation year 1985							
	1	2	3	4	5	6	7	8
6 Manual small	11	32	30	9	25	72	18	7
7 Manual self	12	39	24	24	20	65	90	6
8 Farm	49	79	61	61	75	189	64	140

Table 1.6 is reproduced from Table 2.12 of (Seiyama et al. [76, pp. 46–47]) with the permission granted by authors (copyright holders).

methods proposed to represent, separately or jointly, symmetry and skew-symmetry is presented. The relationships among the models are outlined, and suggestions for strategies of data analysis are provided. As for Chap. 2, applications of the models to a real data set and a software section are provided at the end of the chapter.

In Chap. 4, models and methods of cluster analysis for asymmetric data are covered by considering two main classes: hierarchical and non-hierarchical methods. They are presented and applied to the same small illustrative data set which allows to highlight their different features and capabilities by using, when appropriate, graphical representations of the results. Attention is also paid to the issues of model selection and evaluation which are critical in applications. Although Chap. 4 may be followed independently of the other chapters, some methods rely on the decomposition of the proximities into symmetry and skew-symmetry and the reader can benefit from reading some sections of Chap. 3. Finally, different methods are applied to two real datasets, and the results are compared and discussed. The methods presented can be estimated by complex algorithms which are not available in the standard software. Programs should be requested to the authors of the methods.

In Chap. 5, the focus is on data with three or more ways (*multiway data*), and models appropriate for this kind of data (*multiway models*). Models presented are generally three-way extensions of two-way models covered in Chaps. 2 and 3, which the reader is advised to read ahead to better follow the discussion in Chap. 5. After a general description of three-way data, models for both two-mode and one-mode three-way data are presented. Specifically, different models for two-mode data are distinguished and discussed to represent the asymmetry by the relationships among either objects or dimensions. Finally, an application to real data is presented, where an asymmetric three-way model is compared to a symmetric one, and indications concerning software are also provided.

References

1. Shepard, R. N. (1957). Stimulus and response generalization: A stochastic model relating generalization to distance in psychological space. *Psychometrika, 22*, 325–345.
2. Saito, T., & Yadohisa, H. (2005). *Data analysis of asymmetric structures. Advanced approaches in computational statistics*. Marcel Dekker.

3. Kruskal, J. B. (1964). Multidimensional scaling by optimizing goodness of fit to a nonmetric hypothesis. *Psychometrika, 29*, 1–29.
4. Carroll, J. D., & Arabie, P. (1980). Multidimensional scaling. *Annual Review of Psychology, 31*, 607–649.
5. Young, F. W. (1975). An asymmetric Euclidean model for multi-process asymmetric data. In *Proceedings of the US-Japan Seminar on the Theory, Methods and Applications of Multidimensional Scaling and Related Techniques* (pp. 79–88). University of California.
6. Young, F. W. (1987). *Multidimensional scaling: History, theory, and applications*. R. M. Hamer (Ed.). Lawrence Erlbaum Associates Inc.
7. De Leeuw, J., & Heiser, W. J. (1982). Theory of multidimensional scaling. In P. R. Krishnaiah & L. N. Kanal (Eds.), *Handbook of statistics* (Vol. 2, pp. 285–316). North Holland.
8. Tversky, A. (1977). Features of similarity. *Psychological Review, 84*, 327–352.
9. Tversky, A., & Gati, I. (1978). Studies of similarity. In E. Rosch, & B. Lloyd (Eds.), *Cognition and categorization* (pp. 79–98). Lawrence Erlbaum Associates Inc.
10. Krumhansl, C. L. (1978). Concerning the applicability of geometric models to similarity data: The interrelationship between similarity and spatial density. *Psychological Review, 85*, 445–463.
11. Holman, E. W. (1979). Monotonic models for asymmetric proximities. *Journal of Mathematical Psychology, 20*, 1–15.
12. Nosofsky, R. M. (1991). Stimulus bias, asymmetric similarity, and classification. *Cognitive Psychology, 23*, 94–140.
13. Weeks, D. G., & Bentler, P. M. (1982). Restricted multidimensional scaling models for asymmetric proximities. *Psychometrika, 47*, 201–207.
14. Wilderjans, T. F., Depril, D., & Van Mechelen, I. (2013). Additive biclustering: A comparison of one new and two existing ALS algorithms. *Journal of Classification, 30*, 56–74.
15. Levin, J., & Brown, M. (1979). Scaling a conditional proximity matrix to symmetry. *Psychometrika, 44*(2), 239–243.
16. Bishop, Y. M., Fienberg, S. E., & Holland, P. (1975). *Discrete multivariate analysis; Theory and practice*. M.I.T: Press.
17. Harshman, R. A. (1978). Models for analysis of asymmetrical relationships among N objects or stimuli. *Paper presented at the First Joint Meeting of the Psychometric Society and the Society for Mathematical Psychology*. McMaster University.
18. Chino, N. (1978). A graphical technique for representing the asymmetric relationships between N objects. *Behaviormetrika, 5*, 23–40.
19. Gower, J. C. (1977). The analysis of asymmetry and orthogonality. In J. R. Barra et al. (Eds.), *Recent developments in statistics* (pp. 109–123). North Holland.
20. Constantine, A. G., & Gower, J. C. (1978). Graphical representation of asymmetric matrices. *Applied Statistics, 3*, 297–304.
21. Cox, T. F., & Cox, M. A. A. (2001). *Multidimensional scaling*. CRC/Chapman and Hall.
22. Kruskal, J. B., & Carmone, F. J. Jr. (1971). How to use M-D-SCAL (Version 5M) and other useful information. *Multidimensional scaling program package of Bell Laboratories*, Bell Laboratories.
23. Gower, J. C. (2018). Skew symmetry in retrospect. *Advances in Data Analysis and Classification, 12*, 33–41.
24. Tobler, W. R. (1976). Spatial interaction patterns. *Journal of Environmental Systems, 6*, 271–301.
25. Tobler, W. R. (1979). Estimation of attractivities from interactions. *Environment and Planning A, 11*(2), 121–127.
26. Constantine, A. G., & Gower, J. C. (1982). Models for the analysis of interregional migration. *Environment and Planning (A), 14*, 477–497.
27. Escoufier, Y., & Grorud, A. (1980). Analyse factorielle des matrices carrees non symetriques, In E. Diday et al. (Eds.), *Data analysis and informatics* (pp. 263–276). North-Holland
28. Chino, N., & Shiraiwa, K. (1993). Geometrical structures of some non-distance models for asymmetric MDS. *Behaviormetrika, 20*, 35–47.

29. Lance, G. N., & Williams, W. T. (1966). A generalized sorting strategy for computer classifications. *Nature, 212*, 218.
30. Lance, G. N., & Williams, W. T. (1967). A general theory of classificatory sorting strategies: 1. *Hierarchical systems. The Computer Journal, 9*, 373–380.
31. Hubert, L. J. (1973). Min and max hierarchical clustering using asymmetric similarity measures. *Psychometrika, 38*, 63–72.
32. Johnson, S. C. (1967). Hierarchical clustering schemes. *Psychometrika, 32*, 241–254.
33. Fujiwara, H. (1980). Hitaisho sokudo to toshitsusei keisuu o mochiita kurasuta bunsekiho [Methods for cluster analysis using asymmetric measures and homogeneity coefficient]. *Kodo Keiryogaku [Japanese Journal of Behaviormetrics], 7*(2), 12–21. (in Japanese).
34. Okada, A., & Iwamoto, T. (1995). Hitaisho kurasuta bunnsekihou niyoru daigakushinngaku niokeru todoufukennkann no kanren no bunseki [An asymmetric cluster analysis study on university enrolment flow among Japanese prefectures]. Riron to Houhou [*Sociological Theory and Methods*], *10*, 1–13 (in Japanese).
35. Okada, A., & Iwamoto, T. (1996). University enrollment flow among the Japanese prefectures: A comparison before and after the Joint First Stage Achievement Test by asymmetric cluster analysis. *Behaviormetrika, 23*, 169–185.
36. Yadohisa, H. (2002). Formulation of asymmetric agglomerative hierarchical clustering and graphical representation of its result. *Bulletin of The Computational Statistics Society of Japan, 15*, 309–316. (in Japanese).
37. Takeuchi, A., Saito, T., & Yadohisa, H. (2007). Asymmetric agglomerative hierarchical clustering algorithms and their evaluations. *Journal of Classification, 24*, 123–143.
38. Ozawa, K. (1983). Classic: A hierarchical clustering algorithm based on asymmetric similarities. *Pattern Recognition, 16*, 201–211.
39. Brossier, G. (1982). Classification hiérarchique à partir de matrices carrées non symétriques. *Statistique et Analyse des Données, 7*, 22–40.
40. Olszewski, D. (2011). Asymmetric k-means algorithm. In A. Dovnikar, U. Lotrič, & B. Ster (Eds.), International conference on adaptive and natural computing algorithms (ICANNGA 2011) part II, *Lecture notes in computer science* (Vol. 6594, pp. 1–10). Springer.
41. Olszewski, D. (2012). K-means clustering of asymmetric data. In E. Corchado et al. (Eds.), Hybrid artificial intelligent systems 2012, part I, *Lecture notes in computer science* (Vol. 7208, pp. 243–254). Springer.
42. Olszewski, D., & Šter, B. (2014). Asymmetric clustering using the alpha-beta divergence. *Pattern Recognition, 47*, 2031–2041.
43. Okada, A., & Yokoyama, S. (2015). Asymmetric CLUster analysis based on SKEW-Symmetry: ACLUSKEW. In I. Morlini, T. Minerva, & M. Vichi (Eds.), Advances in statistical models for data analysis. *Studies in classification, data analysis, and knowledge organization* (pp. 191–199). Springer.
44. Okada, A., & Imaizumi, T. (2007). Multidimensional scaling of asymmetric proximities with a dominance point. In D. Baier et al. (Eds.), *Advances in data analysis* (pp. 307–318). Springer.
45. Vicari, D. (2014). Classification of asymmetric proximity data. *Journal of Classification, 31*, 386–420.
46. Vicari, D. (2018). CLUSKEXT: CLUstering model for SKew-symmetric data including EXTernal information. *Advances in Data Analysis and Classification, 12*, 43–64.
47. Vicari, D. (2020). Modelling asymmetric exchanges between clusters. In T. Imaizumi A. Nakayama & S. Yokoyama (Eds.), Advanced studies in behaviormetrics and data science. *Behaviormetrics: Quantitative approaches to human behavior* (pp. 297–313). Springer Nature.
48. Kruskal, J. B., & Wish, M. (1978). *Multidimensional scaling*. Sage.
49. Borg, I., & Groenen, P. J. F. (2005). *Modern multidimensional scaling. Theory and applications* (2nd edn.). Springer.
50. Takane, Y., Jung, S., & Oshima-Takane, Y. (2009). Multidimensional scaling. In R. E. Millsap & A. Maydeu-Olivares (Eds.), *The Sage handbook of quantitative methods in psychology* (pp. 219–242). Sage Publications.
51. Everitt, B. S., & Rabe-Hesketh, S. (1997). *The analysis of proximity data*. Arnold.

52. Gower, J. C., & Legendre, P. (1986). Metric and Euclidean properties of dissimilarity coefficients. *Journal of Classification, 3*, 5–48.
53. Gower, J. C. (2008). Asymmetry analysis: The place of models. In K. Shigemasu & A. Okada (Eds.), *New trends in psychometrics* (pp. 79–86). Universal Academy Press.
54. Hartigan, J. A. (1967). Representation of similarity matrices by trees. *Journal of the American Statistical Association, 62*, 1140–1158
55. Chaturvedi, A., & Carroll, J. D. (2006). CLUSCALE ("CLUstering and multidimensional SCAL[E]ing"): A three-way hybrid model incorporating overlapping clustering and multidimensional scaling structure. *Journal of Classification, 23*, 269–299.
56. Shepard, R. N. (1974). Representation of structure in similarity data: Problems and prospects. *Psychometrika, 39*, 373–421.
57. Kendall, M. G. (1970). *Rank correlation methods* (4th ed.). Griffin.
58. David, H. A. (1988). *The method of paired comparisons.* Griffin.
59. Zielman, B., & Heiser, W. J. (1993). Analysis of asymmetry by a slide-vector. *Psychometrika, 58*, 101–114.
60. Groenen, P. J. F., & Heiser, W. J. (1996). The tunneling method for global optimization in multidimensional scaling. *Psychometrika, 61*(3), 529–550.
61. Tobler, W. R., & Wineberg, S. (1971). A Cappadocian speculation. *Nature, 231*(5297), 39–42.
62. Mair, P., Groenen, P. J. F., & De Leeuw, J. (2020). More on multidimensional scaling and unfolding in R: Smacof Version 2, https://cran.r-project.org/web/packages/smacof/vignettes/smacof.pdf
63. Young, G., & Householder, A. S. (1938). Discussion of a set of points in terms of their mutual distances. *Psychometrika, 3*, 19–22.
64. Torgerson, W. S. (1952). Multidimensional scaling: I. Theory and method. *Psychometrika, 17*, 401–419.
65. Harshmann, R. A., Green, P. E., Wind, Y., & Lundy, M. E. (1982). A model for the analysis of asymmetric data in marketing research. *Marketing Science, 1*, 204–242.
66. DeSarbo, W. S. (1982). GENNCLUS: New models for general nonhierarchical clustering analysis. *Psychometrika, 47*(4), 449–475.
67. Rocci, R., & Bove, G. (2002). Rotation techniques in asymmetric multidimensional scaling. *Journal of Computational and Graphical Statistics, 11*, 405–419.
68. Heiser, W. J., & Busing, F. M. T. A. (2004). Multidimensional scaling and unfolding of symmetric and asymmetric proximity relations. In D. Kaplan (Ed.), *The Sage Handbook of Quantitative Methodology for the Social Sciences* (pp. 25–48). Sage Publications.
69. Caussinus, H., & de Falguerolles, A. (1987). Tableaux carrés: Modélisation et methods factorielles. *Revue de Statistique Appliquée, 35*, 35–52.
70. Krackhardt, D. (1987). Cognitive social structures. *Social Networks, 9*, 109–134.
71. Okada, A. (2011). Centrality of asymmetric social network: Singular value decomposition, conjoint measurement, and asymmetric multidimensional scaling. In S. Ingrassia, R. Rocci, & M.Vichi (Eds.), *New perspectives in statistical modeling and data analysis* (pp. 219–227). Springer.
72. Wasserman, S., & Faust, K. (1994). *Social network analysis: Methods and applications.* Cambridge University Press.
73. Bove, G. (2014). Asymmetries in organizational structures. In D. Vicari, A. Okada, G. Ragozini, & C. Weihs (Eds.), *Analysis and modeling of complex data in behavioral and social sciences* (pp. 65–72). Springer-Verlag.
74. Cailliez, F., & Pages, J. P. (1976). *Introduction à l'analyse des données.* SMASH.
75. European Commission. (2006). *Eurobarometer: Europeans and their languages. Special Eurobarometer 243, Public Opinion Analysis Sector of the European Commission, Brussels.*
76. Seiyama, K., Naoi, A., Sato, Y., Tsuzuki, K., & Kojima, H. (1990). Gendai Nipponn no kaisoukouzou to sono suusei [Stratification structure of contemporary Japan and the trend]. In A. Naoi & K. Seiyama (Eds.), Gendai Nippon no Kaisoukouzou Vol. 1, Shakaikaisou no kouzou to katei [*Social stratification in contemporary Japan, Vol. 1, Structure and process of social stratification*]. Tokyo University Press (pp. 15–50).
77. Okada, A., & Imaizumi, T. (1997). Asymmetric multidimensional scaling of two-mode three-way proximities. *Journal of Classification, 14*, 195–224.

Chapter 2
Methods for Direct Representation of Asymmetry

2.1 Preliminaries

Graphical representation of rectangular data matrices (e.g. subjects by variables) is largely known and can be a convenient starting point to approach the general problem of direct representation of asymmetry in proximity data.

Let us suppose a questionnaire was administered to a group of students in upper secondary school, and three students s_1, s_2 and s_3 in the group have to be compared with respect to maths grade (values in the interval 1/10) and maths self-concept (standardized scale index, values in the interval –2.2/2.3). In Table 2.1, the values of the two variables are reported only for the three students to be compared.

In general, in a rectangular data matrix as the one in Table 2.1, it is interesting to analyse relationships between rows, relationships between columns and relationships between rows and columns. Graphical representations can help these analyses.

A common approach to the analysis of relationships between rows is to compute dissimilarities between pairs of students by the Euclidean distance. In this case, for instance, the dissimilarity for the students s_1 and s_2 is given by

$$\delta_{12} = \sqrt{(9-4)^2 + (1.9-0.1)^2} = \sqrt{5^2 + 1.8^2} = \sqrt{25 + 3.24} = 5.31$$

Dissimilarities δ_{13} and δ_{23} for the other two pairs of students can be computed in a similar way, and the three dissimilarities can be collected in the following dissimilarity matrix Δ

$$\Delta = \begin{pmatrix} \delta_{11} & \delta_{12} & \delta_{13} \\ \delta_{21} & \delta_{22} & \delta_{23} \\ \delta_{31} & \delta_{32} & \delta_{33} \end{pmatrix} = \begin{pmatrix} 0 & 5.31 & 3.1 \\ 5.31 & 0 & 2.2 \\ 3.1 & 2.2 & 0 \end{pmatrix}$$

G. Bove et al., *Methods for the Analysis of Asymmetric Proximity Data*, Behaviormetrics: Quantitative Approaches to Human Behavior 7,
https://doi.org/10.1007/978-981-16-3172-6_2

Table 2.1 Small example of rectangular data matrix

Students	Maths grade	Maths self-concept
s_1	9	1.9
s_2	4	0.1
s_3	6	1.0

where $\delta_{12} = \delta_{21} = 5.31$, $\delta_{13} = \delta_{31} = 3.1$ and $\delta_{23} = \delta_{32} = 2.2$ for the property of symmetry of the dissimilarities, and $\delta_{11} = \delta_{22} = \delta_{33} = 0$ because the dissimilarity of any student from himself is null. Matrix Δ is an example of symmetric proximity matrix.

A direct representation of the entries of matrix Δ can be easily obtained by a scatter plot, where the horizontal and the vertical axes indicate maths self-concept and maths grade, respectively.

The three points in the diagram 2.1 represent the three students with coordinates given by the values of the two variables.

The three geometrical distances d_{12}, d_{13} and d_{23} in the diagram perfectly reproduce the corresponding dissimilarities, so for each pair (i, j), it holds:

$$\delta_{ij} = d_{ij} \ .$$

A large dissimilarity between s_1 and s_2 is easily observed, with student s_1 positioned in the highest levels for both variables and with s_3 in an intermediate position.

The relationships between columns in a rectangular matrix are usually analysed by association measures (which are examples of bipolar proximities measuring similarities). In Table 2.1, the proximity between maths grade and maths self-concept can be measured by their covariance or Pearson correlation coefficient

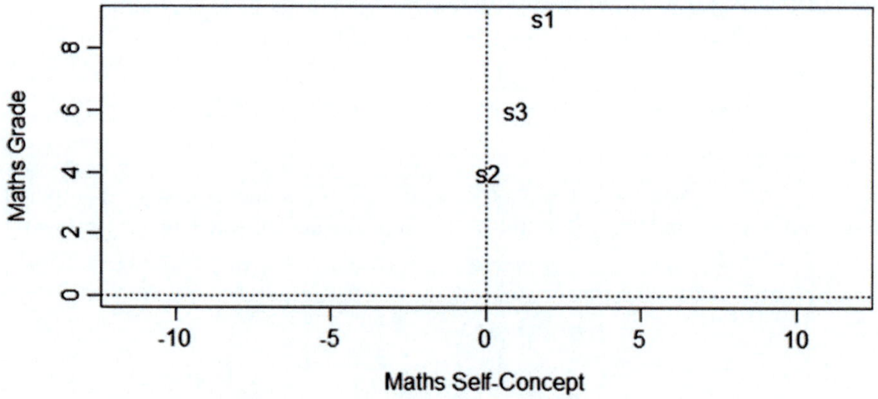

Fig. 2.1 Scatter plot of data in Table 2.1

$$\omega_{12}^{\text{sim}} = cov(\text{M. Grade, M.Self Concept}) = 1.5$$

or

$$\omega_{12}^{\text{sim}} = cor(\text{M. Grade, M. Self Concept}) = 0.99$$

where ω^{sim} indicates a proximity measuring similarity, and *cov* and *cor* denote the covariance and the correlation between the variables in parentheses, respectively. The very high value of the correlation shows a strong concordance between maths self-concept and maths grade, which indicates that high (low) levels of maths self-concept are associated with high (low) levels of maths grade on average.

As for the dissimilarities, covariances or correlations can be collected in a matrix

$$\mathbf{\Omega} = \begin{pmatrix} \omega_{11}^{\text{sim}} & \omega_{12}^{\text{sim}} \\ \omega_{21}^{\text{sim}} & \omega_{22}^{\text{sim}} \end{pmatrix} = \begin{pmatrix} 4.2 & 1.5 \\ 1.5 & 0.54 \end{pmatrix} = \begin{pmatrix} \sigma_{11} & \sigma_{12} \\ \sigma_{21} & \sigma_{22} \end{pmatrix} = \Sigma$$

or

$$\mathbf{\Omega} = \begin{pmatrix} \omega_{11}^{\text{sim}} & \omega_{12}^{\text{sim}} \\ \omega_{21}^{\text{sim}} & \omega_{22}^{\text{sim}} \end{pmatrix} = \begin{pmatrix} 1 & 0.99 \\ 0.99 & 1 \end{pmatrix} = \begin{pmatrix} r_{11} & r_{12} \\ r_{21} & r_{22} \end{pmatrix} = \mathbf{R}$$

where $\mathbf{\Sigma}$ is the covariance matrix (diagonal entries σ_{11} and σ_{22} are the variances of the two variables), and $\mathbf{R}$ is the correlation matrix between maths grade and maths self-concept (diagonal entries are self-correlations). Both matrices are actually particular types of symmetric proximity matrices.

Both the covariance and the correlation could be evaluated in the scatter plot of Fig. 2.1 by examining the degree of linear alignment of the points. However, there is a more convenient way to represent directly covariances and correlations in a separate diagram. Firstly, we perform separately the mean centring of the two variables, by subtracting the column mean from each value (the results are provided in Table 2.2).

Now, it is possible to represent the two variables in a three-dimensional scatter plot, where each axis corresponds to a student, and each variable is represented by a point with coordinates given by the values of the corresponding columns in Table 2.2. Thus, the column vector $\mathbf{c}_1 = (2.66\ -2.33\ -0.33)'$ corresponds to the centred maths grade variable, and the column vector $\mathbf{c}_2 = (0.9\ -0.9\ 0)'$ corresponds to the centred maths self-concept variable. The two variables are represented in the diagram in Fig. 2.2.

Table 2.2 Mean centring of the variables in Table 2.1

Students	C-maths_grade	C-maths_self-concept
s_1	2.66	0.9
s_2	−2.33	−0.9
s_3	−0.33	0.0

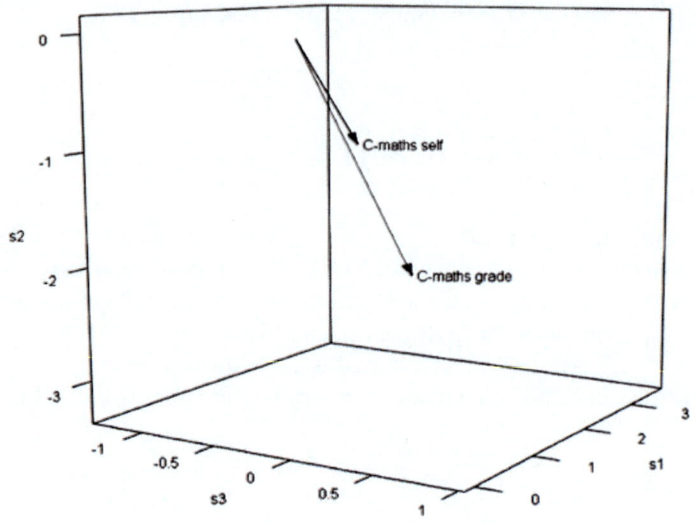

Fig. 2.2 Scatter diagram of variables in three-dimensional space

The inner product between the two vectors is the real number defined as:

$$\mathbf{c}_1'\mathbf{c}_2 = \begin{pmatrix} 2.66 & -2.33 & -0.33 \end{pmatrix} \begin{pmatrix} 0.9 \\ -0.9 \\ 0.0 \end{pmatrix}$$
$$= (2.66 \times 0.9 + 2.33 \times 0.9 - 0.33 \times 0.0) = 4.5$$

Geometrically, the inner product between two vectors (say $\mathbf{c}_1$, $\mathbf{c}_2$) can be visually detected as the product of the length of $\mathbf{c}_1$ ($\mathbf{c}_2$) by the length of the orthogonal projection of $\mathbf{c}_2$ ($\mathbf{c}_1$) on $\mathbf{c}_1$ ($\mathbf{c}_2$). So, an inner product increases as much as the width of the angle between the two vectors decreases.

For the interpretation of Fig. 2.2, it is worth to notice the following relation between the covariance of the variables maths grade and maths self-concept and the inner product $\mathbf{c}_1'\mathbf{c}_2$ of the corresponding columns in Table 2.2

$$\omega_{12}^{\text{sim}} = cov(\text{M. Grade, M. Self Concept}) = \frac{1}{3} \times (\mathbf{c}_1'\mathbf{c}_2).$$

The previous result is general and shows that the covariance between two variables is proportional (according to the scale $1/n$, where n is the number of entries in the vector) to the inner product between the corresponding mean centred vectors. A corollary is that the standard deviation of a variable is proportional (according to the scale $1/\sqrt{n}$) to the *Euclidean norm* (square root of the self-inner product, i.e. the square root of the sum of the squares of the components of the vector—or *length of the vector*) of the corresponding centred vector. At this point, it is easy to show that

$$\omega_{12}^{\text{sim}} = cor(\text{M. Grade, M. Self Concept})$$
$$= \frac{cov(\text{M. Grade, M. Self Concept})}{\sigma_{MG}\sigma_{MSC}}$$
$$= \frac{\mathbf{c}_1^{'}\mathbf{c}_2}{\|\mathbf{c}_1\| \times \|\mathbf{c}_2\|}$$
$$= cos(\mathbf{c}_1, \mathbf{c}_2)$$

where σ denotes the standard deviation, $\|.\|$ denotes the Euclidean norm of the vector, and $cos(\mathbf{c}_1, \mathbf{c}_2)$ is the cosine of the angle the vectors $\mathbf{c}_1, \mathbf{c}_2$ form with the origin of the space.

The previous results allow to enrich the interpretation of Fig. 2.2, where the two vectors representing the centred maths grade and maths self-concept are almost collinear, showing that the value of the covariance between the two variables is positive and large, and the value of the correlation is close to its maximum value 1.

As in the case of the students, where we could interpret the scatter plot as a way to represent directly the entries of a dissimilarity matrix, also in this case, we can consider the three-dimensional scatter diagram as a way to represent directly the entries of a proximity matrix measuring similarity (covariance or correlation matrix).

The multidimensional case

Both the scatter diagrams above were easily obtained because the rectangular data matrix in Table 1.2 has a number of rows and a number of columns not larger than three. Actually, real applications deal with rectangular matrices with large number of rows and columns, so that distances and inner products are defined in many dimensions (*multidimensional spaces*), and proximities cannot be represented exactly in two or three dimensions. To face the problem, *reduction methods* (e.g. techniques of multidimensional scaling or principal component analysis) allow to build approximated representations of proximities between rows or columns in a low number of dimensions, usually two or three dimensions (for a review see, e.g. Gower et al. [1]). The interpretation of the diagrams works in a way similar to the small example presented. A consequence is that the relations between the proximities and the geometrical entities chosen for their graphical representation (*geometrical models*) are approximated. Other reasons for including an error component are that proximities typically contain a certain amount of measurement imprecision, unreliability, sampling effects and so on.

In this case, the relations are reformulated, respectively:

$$\delta_{ij} = d_{ij}(\boldsymbol{x}_i, \boldsymbol{x}_j) + \varepsilon_{ij} \quad (i, j = 1, 2, \ldots, n)$$

$$\omega_{ij}^{\text{sim}} = b_{ij}(\boldsymbol{x}_i, \boldsymbol{x}_j) + \varepsilon_{ij}$$

where ω_{ij}^{sim} is a symmetric proximity measuring similarity, n is the number of rows (columns) of the proximity matrix, d_{ij} denotes a distance, b_{ij} denotes an inner product, $\boldsymbol{x}_i = (x_{i1}, x_{i2} \ldots, x_{ir})'$, $\boldsymbol{x}_j = \left(x_{j1}, x_{j2} \ldots, x_{jr}\right)'$ are column vectors of coordinates in r dimensions to be estimated by the reduction method, for row i and row j in the first relation, for column i and column j in the second relation, respectively, and ε_{ij} is the error component. When proximities are approximated in r dimensions, the most common choices for the distance and the inner product are

$$d_{ij}(\boldsymbol{x}_i, \boldsymbol{x}_j) = \sqrt{\sum_{s=1}^{r} (x_{is} - x_{js})^2} = d_{ji}(\boldsymbol{x}_j, \boldsymbol{x}_i)$$

$$b_{ij}(\boldsymbol{x}_i, \boldsymbol{x}_j) = (\boldsymbol{x}_i'\boldsymbol{x}_j) = \sum_{s=1}^{r} x_{is}x_{js} = b_{ji}(\boldsymbol{x}_j, \boldsymbol{x}_i)$$

where x_{is} and x_{js} are the coordinates of row (column) i and row (column) j on dimension s, respectively, and the symmetric property of both functions is pointed out $(d_{ij} = d_{ji}, b_{ij} = b_{ji})$.

Finally, we remark that *biplot* methods to build approximated representations of the relationships between rows and columns of a rectangular matrix in a low number of dimensions were proposed by Gabriel [2] and other authors (for a review see, e.g. Gower et al. [1]; Greenacre [3]). The prefix "*bi*" in the word biplot refers to the fact that in these approaches, two different sets of points (corresponding to subjects and variables in our example) are depicted jointly in the same diagram (usually a plane). The inner products in the diagram represent directly the entries of the rectangular data matrix (possibly transformed by centring variables) rather than covariances or correlations between variables. According to the chosen normalization of the coordinates, dissimilarities between rows or associations between columns could also be detected in the diagram. In the next section, details will be provided on such approaches when applied to proximity data (including a brief description of a general method to obtain coordinates for rows and columns of a data matrix based on the *singular value decomposition—SVD*). In this section, we provide only the result of the application of the biplot method to our small example in Fig. 2.3.

In order to make the visualization of the inner products easy, in the diagram two vectors are depicted joining the origin with the variable points. Each of these vectors defines a *biplot axis* (oriented according to vector direction) which subject points s_1, s_2 and s_3 can be perpendicularly projected onto (in passing, we notice that, equivalently, biplot axes could be defined by vectors joining the origin with each subject point which variable points could be projected onto). According to the inner product definition, the length of each projection multiplied by the length of the vector generating the axis provides the corresponding entry in the data matrix. Since all the projections on the same axis are multiplied by the same factor given by the length of the generating vector, the simple visual comparison of the projections on the same axis provides information on the ranking of the values the subjects have

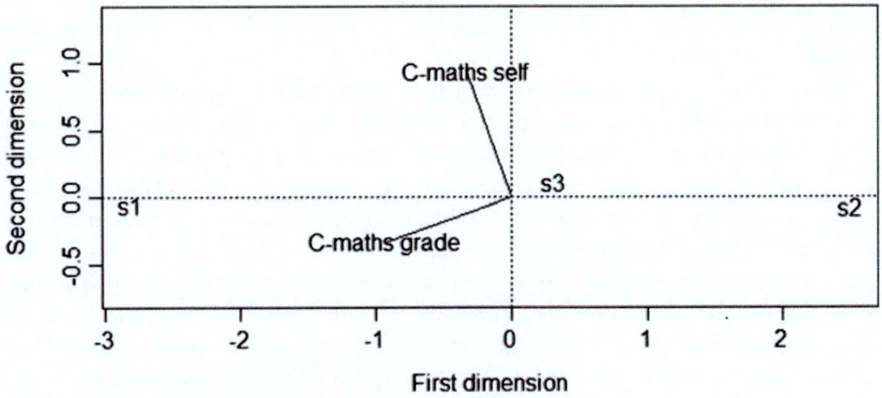

Fig. 2.3 Biplot representation of data in Table 2.2

on the variable associated with the axis. So, in Fig. 2.3, we can see, for instance, that subject s_1 has the projection with the largest length on the positive side of the axis generated by the maths self-concept vector, so s_1 has the highest positive value on the variable (maths self-concept column in Table 2.2). Subject s_2 has the projection with the largest length on the negative side of the axis generated by the maths self-concept vector, so s_2 has the smallest value on the variable, and subject s_3 is positioned between s_1 and s_2. Analogously, we can compare the ranking of the subject values on maths grade variable projecting the corresponding points on the biplot axis of the variable. Besides, we have chosen a normalization of the coordinates in Fig. 2.3 that allows also for the analysis of the dissimilarities between subjects by the corresponding distances (variables centring does not affect the dissimilarities between subjects). Figure 2.3 provides the same information concerning subject dissimilarities depicted in Fig. 2.1.

Towards proximity analysis

After the considerations regarding graphical representation of rectangular matrices in the previous section, we are now ready to present the methodological problem addressed later in this chapter. Suppose we are involved in an experiment where each student in a class is asked to compare their classmates in pairs and say who is the best in math (or say if they are equivalent). How often student *i* is rated better in maths than student *j* can be assumed to be a measure of their dissimilarity.

So, an *asymmetric proximity* matrix can be built where students in the class correspond to rows and columns, and entries are the frequencies observed in the experiment (an example will be provided in Chap. 2, Table 2.4). As for the dissimilarity matrix between the three students of the previous small example, also in this experiment, we may be interested in analysing student differences in maths by depicting proximities in a graphical representation.

A second example that can be considered regards relationships between variables, where the correlations between the same variables in two different occasions are known, but the original subject × variable matrices at the different occasions are not available or unknown.

The correlations measure the similarity between variables, and the resulting proximity matrix is asymmetric because correlation between variable i at occasion 1 and variable j at occasion 2 is different from the correlation between variable j at occasion 1 and variable i at occasion 2. As for the small proximity matrix between maths grade and maths self-concept, we are interested in analysing correlations by depicting their values in a graphical representation.

Compared to the small example, there are two differences: (1) the proximities are asymmetric, and (2) since the underlying information of the rectangular matrix which originated the proximities is not available, the graphical representation of the students and/or variables need to be built directly from the proximities themselves.

Several methods are available to represent directly all elements ω_{ij} of an asymmetric proximity matrix $\boldsymbol{\Omega}$. A distinction between models depends on whether the elements of $\boldsymbol{\Omega}$ are thought of as some kind of *inner products* or of *pseudo-distances*. Bilinear models are usually considered for the first case and distance-like models for the second. Inner product and distance models are suitably modified by increasing the number of parameters to represent the asymmetry between ω_{ij} and ω_{ji}. In the following two sections, some methods based on bilinear and distance-like models are presented, and applications will be provided in the last section of the chapter to emphasize their capabilities and effectiveness. Model formulation will be specified for the data to be processed such that data matrices to be represented are coherent with the different models presented.

2.2 Bilinear Methods

Any proximity (or rectangular) matrix may be represented by associating a vector to each row and another vector to each column so that the entries of the matrix are the inner products of the vectors associated to the corresponding rows and columns. Let us consider a simple example concerning import/export flows between three countries A, B and C.

In Table 2.3, 33 is the volume of export from country A directed to country B, and the diagonal entry 36 is the volume exchanged within country A (internal flow). Similar interpretations hold for the other entries. Country A has the highest export levels, and country C has high import levels compared to the export, while for country B, total exports and total imports are more equilibrated. The proximity matrix $\boldsymbol{\Omega}$

Table 2.3 Import/export flows between three countries

Countries	A	B	C
A	36	33	39
B	16	26	14
C	7	5	8

based on Table 2.3 is

$$\boldsymbol{\Omega} = \begin{pmatrix} 36 & 33 & 39 \\ 16 & 26 & 14 \\ 7 & 5 & 8 \end{pmatrix},$$

and to prove that each entry can be considered as an inner product between a row vector and a column vector, the following two matrices $\mathbf{C}_0$ and $\mathbf{C}_1$ can be defined:

$$\mathbf{C}_0 = \begin{pmatrix} 3 & 6 \\ -2 & 4 \\ 1 & 1 \end{pmatrix}$$

$$\mathbf{C}_1 = \begin{pmatrix} 2 & 5 \\ -1 & 6 \\ 3 & 5 \end{pmatrix}.$$

From the definition of the inner product between vectors given in Sect. 2.1, it is easy to show that the following *factorization* holds:

$$\boldsymbol{\Omega} = \begin{pmatrix} 36 & 33 & 39 \\ 16 & 26 & 14 \\ 7 & 5 & 8 \end{pmatrix} = \begin{pmatrix} 3 & 6 \\ -2 & 4 \\ 1 & 1 \end{pmatrix} \begin{pmatrix} 2 & -1 & 3 \\ 5 & 6 & 5 \end{pmatrix} = \mathbf{C}_0\, \mathbf{C}_1'$$

where, for instance, entry 36 turns out to be the inner product of the first row vector of matrix $\mathbf{C}_0$ and the first column vector of matrix $\mathbf{C}_1'$ (transpose of matrix $\mathbf{C}_1$), that is

$$36 = \begin{pmatrix} 3 & 6 \end{pmatrix} \begin{pmatrix} 2 \\ 5 \end{pmatrix} = (3 \times 2) + (6 \times 5).$$

Matrices $\mathbf{C}_0$ and $\mathbf{C}_1$ have only two columns, so their rows can be assumed as coordinate vectors in two dimensions of the rows and columns of the proximity matrix $\boldsymbol{\Omega}$, respectively. Note that *equivalent factorizations* of matrix $\boldsymbol{\Omega}$ can be obtained with appropriate transformations of matrices $\mathbf{C}_0$ and $\mathbf{C}_1$ (e.g. an equivalent factorization of matrix $\boldsymbol{\Omega}$ is obtained if entries in the two columns of matrix $\mathbf{C}_0$ are multiplied by scale factors λ_1 and λ_2, respectively, and entries in the two columns of matrix $\mathbf{C}_1$ are multiplied by scale factors $1/\lambda_1$ and $1/\lambda_2$, respectively); that is, factorization $\mathbf{C}_0\mathbf{C}_1'$ is not unique. Nevertheless, any equivalent factorization of matrix $\boldsymbol{\Omega}$ by two matrices of dimensions 3×2 is suitable for the geometrical representation of proximities by inner products.

The diagram in Fig. 2.4a provides a geometrical representation of the proximities by inner products corresponding to the factorization $\mathbf{C}_0\mathbf{C}_1'$. Capital letters represent

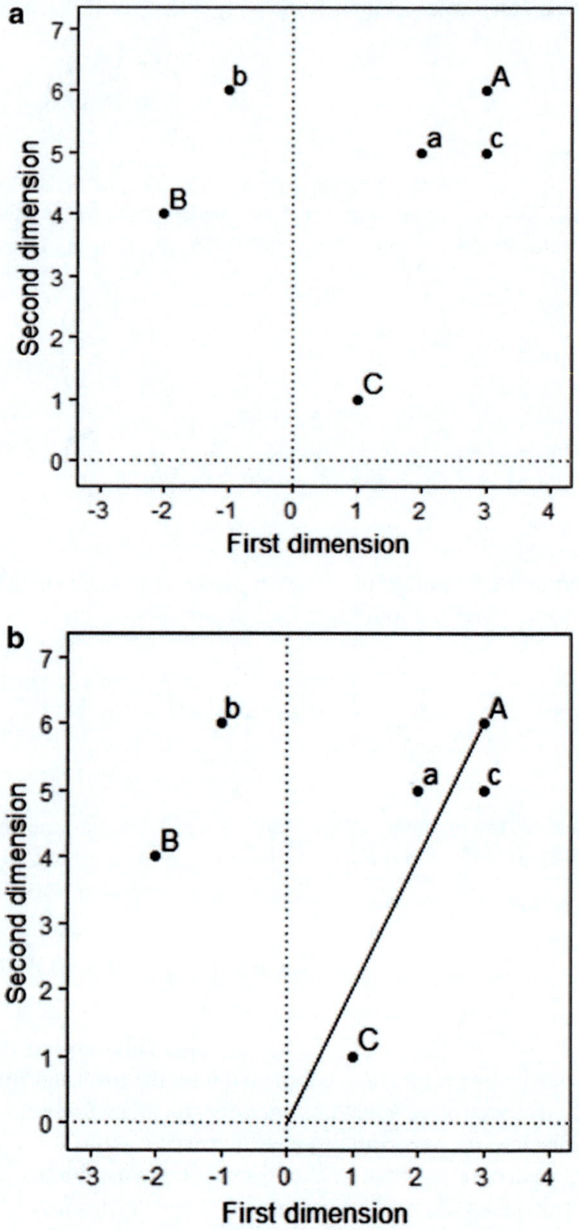

Fig. 2.4 **a** Import/export flows of Table 2.3 represented by matrices $\mathbf{C}_0$ and $\mathbf{C}_1$ **b** ranking of export flows of country A in Table 2.3 represented by the projections of points a, b and c on the vector from the origin to point A

rows (countries of origin of the flow), and small letters represent columns (countries of destination of the flow).

Country A is far from the origin, so the inner products with the other countries tend to be high as the corresponding flows in the first row of Table 2.3. Country C is close to the origin, so the inner products with the other countries are low as the flows in the third row of the table.

As in the biplot diagram of Fig. 2.3, vectors joining the origin with each row (column) point could be drawn in the diagram to define axes which column (row) points can be perpendicularly projected onto. Thus, rankings of export (import) flows related to a country could be analysed through the orthogonal projections of small (capital) letters on the axis defined by the vector from the origin to the capital (small) letter for any country. For instance, if small letters are projected on the axis defined by the vector depicted in Fig. 2.4b, the ranking of the flows from country A to destination countries can be detected. The projection of country c is the longest one, followed by the projection of country a (internal flow) and country b. The same ranking characterizes the flows from country C, but at much lesser extent and with small differences among countries because of the short length of the vector corresponding to capital C. The reverse ranking is observed for country B, with a strong incidence of the internal flow with respect to the two export flows.

Singular value decomposition and bilinear model

Now, we are in the position to generalize the previous factorization to any proximity matrix $\mathbf{\Omega}$. In fact, it can be shown that for any $(n \times n)$ matrix $\mathbf{\Omega}$ of *rank* q (the rank of a matrix is the minimum number of row or column vectors that generate all the rows or columns of the matrix itself exactly through linear combinations), the following general matrix decomposition (named *Singular Value Decomposition—SVD)* holds.

$$\mathbf{\Omega} = \mathbf{G}\mathbf{D}_\alpha\mathbf{V}', \tag{2.1}$$

where $\mathbf{G}$ and $\mathbf{V}$ are $n \times q$ matrices, and $\mathbf{D}_\alpha$ is a $q \times q$ diagonal matrix with positive numbers (named *singular values*) as diagonal entries given in descending order ($\alpha_1 \geq \alpha_2 \geq \ldots \geq \alpha_q > 0$). Furthermore, the columns of $\mathbf{G}$ and $\mathbf{V}$ (named *left* and *right-singular vectors*, respectively) are *orthonormal* (i.e., inner products between each pair of different columns are zero and each column vector has unit length). We remark that the decomposition (2.1) holds also for any rectangular matrix (by setting the number of rows of matrix $\mathbf{V}$ equal to the number of columns of the rectangular matrix).

In scalar notation, the decomposition is

$$\omega_{ij} = \sum_{s=1}^{q} \alpha_s g_{is} v_{js}, \tag{2.2}$$

where g_{is} is the *i*-th entry of the column singular vector *s* of matrix **G**, and v_{js} is the *j*-th entry of the column singular vector *s* of matrix **V**.

Hence, the two matrices of the factorization (say $\mathbf{C}_0$ and $\mathbf{C}_1$) can be obtained in many different ways from the SVD (e.g. $\mathbf{C}_0 = \mathbf{GD}_\alpha$ and $\mathbf{C}_1 = \mathbf{V}$). When $\mathbf{\Omega}$ has a large number of rows and columns, however, the coordinate matrices have usually more than two or three columns, so entries of matrix $\mathbf{\Omega}$ cannot be represented exactly by inner products in two or three dimensions. In order to obtain diagrams as the one in Fig. 2.4a, we need an optimal approximation of the decomposition of $\mathbf{\Omega}$ in a low number of dimensions (hopefully two or three dimensions). A consequence of the approximation is that the relations between the proximities and the inner products will be reformulated incorporating an error component (similarly to Sect. 2.1 for symmetric proximities). The approximated decomposition can be formulated in matrix notation as a *bilinear model*

$$\mathbf{\Omega} = \mathbf{B} + \mathbf{E} = \mathbf{XY}' + \mathbf{E},$$

where **B** is a $(n \times n)$ matrix of inner products with rank r, **X** and **Y** are $(n \times r)$ coordinate matrices ($r \leq q$, where usually $r = 2$ or $r = 3$) to be estimated, and matrix **E** is an $(n \times n)$ matrix of error components. The rows of **X** and **Y** provide coordinate vectors for the n rows and columns of the proximity matrix, respectively, that allow us to represent asymmetric proximities by $2n$ points in a low-dimensional space (usually a plane).

In scalar notation, the *bilinear model* is

$$\omega_{ij} = b_{ij}(\boldsymbol{x}_i, \boldsymbol{y}_j) + \varepsilon_{ij} = \sum_{s=1}^{r} x_{is} y_{js} + \varepsilon_{ij},$$
$$(i, j = 1, 2, \ldots, n) \tag{2.3}$$

where ω_{ij} is the observed proximities (usually assumed to measure similarities), b_{ij} denotes an inner product, $\boldsymbol{x}_i = (x_{i1}, x_{i2} \ldots, x_{ir})'$, $\boldsymbol{y}_j = (y_{j1}, y_{j2} \ldots, y_{jr})'$ are column vectors of coordinates in an *r-dimensional* space, x_{is} and y_{js} are the coordinates of row i and column j on dimension s, respectively, and ε_{ij} is an error component. In a proximity matrix rows and columns correspond to the same set of objects (e.g. countries, stimuli, persons, etc.), which means that each object is represented twice in the *r-dimensional* space (e.g. for $r = 2$, any object is represented twice in the plane, by coordinates (x_{i1}, x_{i2}) as a row point and by coordinates (y_{i1}, y_{i2}) as a column point). The projection of a column (row) point onto the axis generated by a row (column) point vector, multiplied by the length of the row (column) point vector, gives the inner product approximating the corresponding entry of $\mathbf{\Omega}$. Asymmetry is derived by comparing the two inner products corresponding to entries ω_{ij} and ω_{ji}.

Eigenvalue decomposition of a symmetric matrix

In the special case where $\boldsymbol{\Omega}$ is symmetric, the following decomposition (named *Eigenvalue Decomposition*—EVD) holds

$$\boldsymbol{\Omega} = \mathbf{A}\mathbf{D}_{\lambda}\mathbf{A}'$$

where $\mathbf{A}$ is of size $n \times q$ with orthonormal columns (called *eigenvectors*), and $\boldsymbol{D}_{\lambda}$ is a $q \times q$ diagonal matrix with real numbers $\lambda_1 \geq \lambda_2 \geq \cdots .. \geq \lambda_q$ (called *eigenvalues*) as diagonal entries. We remark that when all eigenvalues are nonnegative, SVD and EVD coincide ($\mathbf{G} = \mathbf{V} = \mathbf{A},\ \mathbf{D}_{\alpha} = \mathbf{D}_{\lambda}$). The approximated decomposition of matrix $\boldsymbol{\Omega}$ can be rewritten

$$\boldsymbol{\Omega} = \mathbf{B} + \mathbf{E} = \mathbf{X}\mathbf{X}' + \mathbf{E},$$

and model (2.3) becomes the standard inner product model considered in Sect. 2.1 (i.e. $y_{is} = x_{is}$).

Estimation of matrix B.

A good approximation of matrix $\boldsymbol{\Omega}$ with a matrix $\mathbf{B}$ of rank r is obtained when the error component matrix $\mathbf{E}$ is "small", so that matrix $\mathbf{B}$ resembles $\boldsymbol{\Omega}$ as closely as possible. A simple criterion to evaluate the "size" of the error component is the *least-squares criterion (LS)* that is defined as

$$LS(\boldsymbol{\Omega}, \mathbf{B}) = \|\boldsymbol{\Omega} - \mathbf{B}\|^2 = \|\mathbf{E}\|^2 = \sum_i \sum_j \varepsilon_{ij}^2,$$

where the symbol $\|.\|$ represents the Frobenius norm of a matrix, computed as the square root of the sum of the squares of all the elements.

A general method to estimate matrix $\mathbf{B}$ that minimizes the least-squares criterion is based on the SVD of matrix $\boldsymbol{\Omega}$. An important property is that the best approximation of $\boldsymbol{\Omega}$ with a matrix $\mathbf{B}$ of rank r is

$$\mathbf{B} = \mathbf{G}_r \mathbf{D}_{\alpha_r} \mathbf{V}'_r, \tag{2.4}$$

where $\mathbf{G}_r$ and $\mathbf{V}_r$ are obtained by taking only the first r columns of matrices $\mathbf{G}$ and $\mathbf{V}$, respectively, and matrix $\mathbf{D}_{\alpha_r}$ is the upper left $r \times r$ part of matrix $\mathbf{D}_{\alpha}$.

Quality of approximation

Singular values $\alpha_1, \alpha_2, \ldots, \alpha_r$ are the basic elements for measuring the resemblance between matrices $\boldsymbol{\Omega}$ and $\mathbf{B}$. The measure is based on the concept of *explained variance* (or *Variance Accounted For—VAF*) and uses the following three properties (the symbol SS stands for sum of squares):

$$\mathrm{SS}(\boldsymbol{\Omega}) = \sum_{j=1}^{n}\sum_{j=1}^{n}\omega_{ij}^2 = \sum_{s=1}^{q}\alpha_s^2$$

$$\mathrm{SS}(\mathbf{B}) = \sum_{j=1}^{n}\sum_{j=1}^{n}b_{ij}^2 = \sum_{s=1}^{r}\alpha_s^2$$

$$\mathrm{SS}(\boldsymbol{\Omega} - \mathbf{B}) = \mathrm{SS}(\mathbf{E}) = LS(\boldsymbol{\Omega}, \mathbf{B})$$
$$= \sum_{i=1}^{n}\sum_{j=1}^{n}\varepsilon_{ij}^2 = \sum_{s=r+1}^{q}\alpha_s^2.$$

Hence, the following decomposition holds:

$$\mathrm{SS}(\boldsymbol{\Omega}) = \mathrm{SS}(\mathbf{B}) + \mathrm{SS}(\boldsymbol{\Omega} - \mathbf{B}) = \mathrm{SS}(\mathbf{B}) + \mathrm{SS}(\mathbf{E}),$$

that suggests to measure the quality of the approximation by the following ratio

$$\frac{\mathrm{SS}(\mathbf{B})}{\mathrm{SS}(\boldsymbol{\Omega})}$$

that represents the *proportion of variance of matrix* $\boldsymbol{\Omega}$ *explained by matrix* **B**. The ratio is always positive and not larger than one, so can be expressed as a percentage by multiplying it by 100. In the examples, matrix $\boldsymbol{\Omega}$ is represented exactly in the diagrams so the total variance is fully explained (100% explained variance). The higher the value of the ratio, the more confident we are in using the diagram for the analysis of matrix $\boldsymbol{\Omega}$. The ratio

$$\frac{\mathrm{SS}(\mathbf{E})}{\mathrm{SS}(\boldsymbol{\Omega})}$$

represents the *residual* (or *error* or *unexplained*) *proportion of variance* of matrix $\boldsymbol{\Omega}$ (it can also be expressed in percentage multiplying it by 100), and it is useful to evaluate whether the error term **E** is "small" enough.

Definitions of matrices **X** *and* **Y**

Matrices **X** and **Y** can be obtained in different equivalent ways from the definition $\mathbf{B} = \mathbf{XY}'$ according to how $\mathbf{D}_{\alpha_r}$ is handled (see also Greenacre [3]. Three are the most used choices:

Symmetric assignment

$$\mathbf{X} = \mathbf{G}_r\,\mathbf{D}_{\alpha_r}{}^{1/2}$$

$$\mathbf{Y} = \mathbf{V}_r\,\mathbf{D}_{\alpha_r}{}^{1/2}$$

Left assignment

$$\mathbf{X} = \mathbf{G}_r\mathbf{D}_{\alpha_r}$$

$$\mathbf{Y} = \mathbf{V}_r$$

Right assignment

$$\mathbf{X} = \mathbf{G}_r$$

$$\mathbf{Y} = \mathbf{V}_r\mathbf{D}_{\alpha_r}$$

where the diagonal matrix $\mathbf{D}_{\alpha_r}{}^{1/2}$ has the square root of entries α_i $(i = 1, 2, \ldots, r)$ as diagonal elements. Coordinates obtained by assigning matrix $\mathbf{D}_{\alpha_r}$ completely to the left $\left(\mathbf{G}_r\mathbf{D}_{\alpha_r}\right)$ or to the right $\left(\mathbf{V}_r\mathbf{D}_{\alpha_r}\right)$ are called *principal coordinates*, while coordinates that coincide with one of the two matrices of singular vectors ($\mathbf{G}_r$ or $\mathbf{V}_r$) are called *standard coordinates*. In addition to the interpretations of inner products between rows and columns, diagrams obtained by left assignment allow for the interpretation of the distances between rows of $\mathbf{\Omega}$ in terms of distances between row points, and diagrams obtained by right assignment allow for the interpretation of the distances between columns of $\mathbf{\Omega}$ in terms of distances between column points. On the other hand, the assignment of singular values completely to one side can make the scale of the principal coordinates so large compared to the scale of standard coordinates to render the visual analysis too difficult. This suggests to choose representations with coordinates from the symmetric assignment in many applications. It is worth to notice that the three assignments provide similar diagrams when the singular values are close to one (in these cases, the assignment makes simply a small expansion or contraction of the clouds of points obtained from the standard coordinates). The diagram in Fig. 2.5a provides the representation of the import/export flows of Table 2.3 obtained by the symmetric assignment coordinates in the SVD. A 90° clockwise rotation was applied for an easy comparison with the configuration in Fig. 2.4a.

An important property of the matrices $\mathbf{X}$ and $\mathbf{Y}$ minimizing $LS(\mathbf{\Omega}\ ,\ \mathbf{B})$ is that the solution in r_1 dimensions is contained in the solution in r_2 dimensions for $r_1 < r_2$ (*nested solutions*). So, for instance, the coordinates of the solution in two dimensions are obtained by the coordinates of the solution in three dimensions, by discarding the third dimension.

Software information concerning programs to compute matrices $\mathbf{X}$ and $\mathbf{Y}$ for the three types of assignment, and the R commands to reproduce the diagram in Fig. 2.5a are provided in Sect. 2.6.

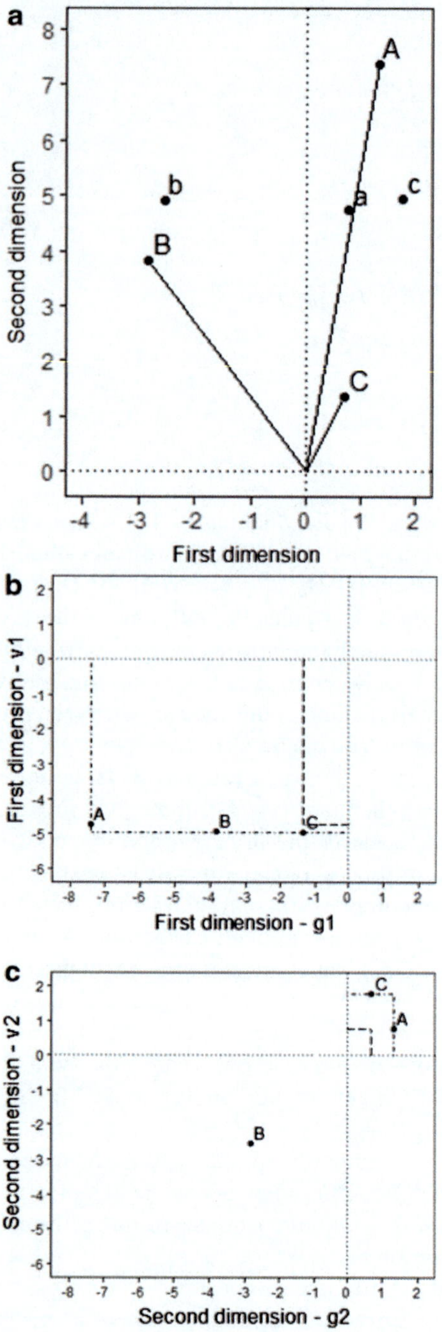

Fig. 2.5 **a** Import/export flows of Table 2.3 represented by the SVD symmetric assignment coordinates, **b** import/export flows of Table 2.3 represented by the Okada–Tsurumi method with symmetric assignment coordinates (dimensions g_1, v_1; rectangles drawn for countries A, C) and **c** import/export flows of Table 2.3 represented by the Okada–Tsurumi method with symmetric assignment coordinates (dimensions g_2, v_2; rectangles drawn for countries A, C)

Okada–Tsurumi method

In the approximation in two dimensions ($r = 2$), a different approach is proposed in Okada and Tsurumi [4] who provide a representation of matrix $\boldsymbol{\Omega}$ in two diagrams with n points each (instead of one diagram with $2n$ points as in the bilinear model). The method proposed can be briefly described by considering the scalar version of the SVD given in (2.2). In the first diagram, n points are displayed with coordinates (g_{i1}, v_{i1}); in the second diagram, n points are displayed with coordinates (g_{i2}, v_{i2}). By recalling the definition of matrix $\mathbf{B}$ in (2.4), for $r = 2$, the approximations b_{ij} of the proximities ω_{ij} can be written

$$b_{ij} = \alpha_1 g_{i1} v_{j1} + \alpha_2 g_{i2} v_{j2}.$$

Thus, in the first of the two diagrams proposed, the absolute value of $g_{i1}v_{j1}$ is the area of a rectangle with base on the first dimension having coordinates $(0, 0)$, $(g_{i1}, 0)$ and height on the second dimension having coordinates $(0, 0)$, $(0, v_{j1})$. Similarly, in the second diagram, the absolute value of $g_{i2}v_{j2}$ is the area of a rectangle with base on the first dimension having coordinates $(0, 0)$, $(g_{i2}, 0)$ and height on the second dimension having coordinates $(0, 0)$, $(0, v_{j2})$. The linear combination of the areas of the two rectangles with coefficients α_1, α_2 provides the approximation of ω_{ij}. In order to evaluate the asymmetry, the approximations of ω_{ij} and ω_{ji} need to be compared. The comparison is made easier by the symmetric assignment of the singular values as in the bilinear model, so the coordinates of any point i in the diagram for dimension s are $\left(\sqrt{\alpha_s} g_{is}, \sqrt{\alpha_s} v_{is}\right)$, for $s = 1,2$.

The diagrams in Fig. 2.5b and c provide the configurations of the import/export flows of Table 2.3 from the Okada–Tsurumi method. The algebraic sum of the areas of the rectangles drawn with dash-dot lines represents the flow from country A to country C (the flow is 39); the algebraic sum of the areas of the rectangles drawn with dashed lines represents the flow from country C to country A (the flow is 7). It turns out that the flow from country A to country C is greater than the flow from country C to country A. The other flows of the table can be compared in the same way, by drawing the corresponding rectangles.

An advantage of the method by Okada–Tsurumi is that rectangle areas are easier than inner products to detect in the diagrams, and that more than two dimensions can be analysed by comparing the corresponding planar diagrams, while a disadvantage is that we need to visualize linear combinations (or sums) of rectangle areas in two diagrams. In Okada–Tsurumi [4] practical indications are provided for diagrams interpretation in an application to brand switching data. The method can be applied by any computer program including routines for the SVD computation and graphical representation. The R commands that reproduce the diagrams in Fig. 2.5b and c are provided in the software Sect. 2.6.

2.3 Distance-like Models

In this section, the main concepts related to the distance models for the analysis of symmetric proximities will be recalled first, because they are important also for the analysis of asymmetric proximities. Then, some methods to deal with the direct representation of asymmetric proximities will be presented.

2.3.1 *Main Concepts on Distance Representation of Symmetric Proximities*

In Sect. 2.1, it was shown that a dissimilarly matrix $\Delta = (\delta_{ij})$ related to n objects can be approximated in an r-dimensional space ($r < q = rank\ (\Delta)$) by the *distance model*

$$\delta_{ij} = d_{ij}\big(\boldsymbol{x}_i, \boldsymbol{x}_j\big) + \varepsilon_{ij} \quad (i, j = 1, 2, \ldots, n),$$

where $d_{ij}\big(\boldsymbol{x}_i, \boldsymbol{x}_j\big)$ denotes a (usually Euclidean) distance, $\boldsymbol{x}_i = \big(x_{i1}, x_{i2} \ldots, x_{ir}\big)'$, $\boldsymbol{x}_j = \big(x_{j1}, x_{j2} \ldots, x_{jr}\big)'$ are column vectors of coordinates in r dimensions, and ε_{ij} is the error component. Coordinate vectors can be collected in a $(n \times r)$ coordinate matrix $\mathbf{X}$. Thus, the set of n objects is represented as a cloud of points in an *r-dimensional* space, so that the distances between points approximate the corresponding observed dissimilarities.

Finding an appropriate value of r and the coordinate matrix $\mathbf{X}$ that "approximate at best" the dissimilarities is one of the main purposes of the techniques of multidimensional scaling (MDS), a class of methods for the analysis of proximity data. The multiplicity of methods available differs in how the discrepancy between observed dissimilarities and fitted distances is measured (*approximation criterion* or *objective function* or *loss function*). Methods optimizing the same approximation criterion can differ with respect to the computational algorithm, but this is a technical aspect that will not be considered.

One of the first methods proposed is called *classical* MDS [5, 6], based on the eigenvalue decomposition of a symmetric matrix (see Sect. 2.2). The idea is simple, the observed dissimilarity matrix is considered as a distance matrix, so the preliminary transformation (1.1) to an inner product matrix is performed. Then, for the inner product matrix, the eigenvalue decomposition is computed $\big(\mathbf{B} = \mathbf{X}\mathbf{X}'\big)$ to obtain a multidimensional representation in r dimensions (for details see, e.g. Borg and Groenen [7]). The coordinate matrix $\mathbf{X}$ shares the property of nested solutions presented in Sect. 2.2 for bilinear models.

The equality relation between dissimilarities and distances can be too stringent for dissimilarities measured at interval or ordinal scale level. In such cases, the measurement level can be taken into account by incorporating a monotonic transformation in the computational algorithm for model estimation.

The distance model can be reformulated by including the monotone transformation f, mapping the proximities ω_{ij} into a set of transformed values $\hat{d}_{ij} = f(\omega_{ij})$, called *disparities* or *pseudo-distances* (labelled *d-hats*)

$$f(\omega_{ij}) = \hat{d}_{ij} = d_{ij}\left(\boldsymbol{x}_i, \boldsymbol{x}_j\right) + \varepsilon_{ij} \quad (i, j = 1, 2, \ldots, n). \tag{2.5}$$

A scaling algorithm has to be able to find the *best monotonic transformation f* of the proximities (also named *optimal scaling*), as well as the coordinate matrix **X** (*object configuration*) which best fits the transformed data. As for bilinear models, to evaluate when the error component is small enough requires an explicit definition of a least-squares approximation criterion. Moreover, a system of weights for the observed proximities can be considered in the least-squares criterion (e.g. to deal with missing observations or to fix certain points with more accuracy than others).

Approximation criteria and optimization

Now, some least-squares criteria that will be widely used hereafter are recalled, pointing out relationships and the main aspects of their optimization.

The "*kernel*" on which most least-squares approximation criteria are based in this context is *raw Stress* σ_r, termed by Kruskal [8, p. 8], given by

$$\begin{aligned}\mathrm{LS}_{wf}(\boldsymbol{\Omega}, \mathbf{D}, \boldsymbol{\theta}) = \sigma_r &= \sum\sum\nolimits_{i<j} w_{ij}\left(f(\omega_{ij}) - d_{ij}\left(\boldsymbol{x}_i, \boldsymbol{x}_j\right)\right)^2 \\ &= \sum\sum\nolimits_{i<j} w_{ij}\left(\hat{d}_{ij} - d_{ij}\left(\boldsymbol{x}_i, \boldsymbol{x}_j\right)\right)^2,\end{aligned} \tag{2.6}$$

where the transformation f depends on the parameter vector $\boldsymbol{\theta}$ (e.g. $\boldsymbol{\theta} = (\alpha, \beta)$ for the linear transformation $\hat{d}_{ij} = \alpha\, \omega_{ij} + \beta$) and w_{ij} are the weights. In matrix notation, the raw stress is

$$LS_{wf}\left(\widehat{\mathbf{D}}, \mathbf{D}\right) = \left\|\widehat{\mathbf{D}} - \mathbf{D}\right\|_{\mathrm{W}}^2$$

where $\widehat{\mathbf{D}} = \left(\hat{d}_{ij}\right)$ is the matrix of the disparities, $\mathbf{D} = (d_{ij}\left(\boldsymbol{x}_i, \boldsymbol{x}_j\right))$ is the matrix of the fitted distances, and $\|.\|_{\mathrm{W}}$ is the weighted norm of a matrix. Note that for a $(n \times n)$ matrix $\mathbf{A} = \left(a_{ij}\right)$ is $\|\mathbf{A}\|_{\mathrm{W}} = \sum_{i=1}^{n} \sum_{j=1}^{n} w_{ij} a_{ij}^2$.

The optimal transformation and the coordinate matrix **X** cannot be determined by SVD or EVD, but they need to be derived by iterative algorithms (named Alternating Least Squares—*ALS* algorithms) that minimize the raw stress (2.6) with respect to both **X** and the transformation f, subject to normalization restrictions for the transformed proximities or the distances. The normalization restrictions are required in order to avoid degenerate solutions (diagrams where all the row and column points coincide, or row and column points are completely separated in the diagram) and to remove the dependence of criterion (2.6) on the scale of the configuration. As a result, the normalized criteria are usually invariant under rotations, translations, reflections and changes of the scales.

An example of approximation criterion incorporating a normalization restriction for disparities is the *Normalized raw Stress* σ_n, given by

$$\sigma_n = \frac{\sum\sum_{i<j} w_{ij}\left(\hat{d}_{ij} - d_{ij}(\boldsymbol{x}_i, \boldsymbol{x}_j)\right)^2}{\sum\sum_{i<j} w_{ij}\hat{d}_{ij}^2}. \tag{2.7}$$

So, σ_n is obtained dividing σ_r in (2.6) by the square of the weighted norm of the disparities. Another example of criterion incorporating a normalization restriction for disparities can be found in Sammon [9].

A widely used approximation criterion incorporating a normalization restriction for distances is

$$Stress - \mathrm{I} = \sigma_1 = \sqrt{\frac{\sum\sum_{i<j} w_{ij}\left(\hat{d}_{ij} - d_{ij}(\boldsymbol{x}_i, \boldsymbol{x}_j)\right)^2}{\sum\sum_{i<j} w_{ij} d_{ij}^2(\boldsymbol{x}_i, \boldsymbol{x}_j)}}, \tag{2.8}$$

where the reason for adding the square root is to make possible a comparison with the error distribution (analogous to the choice of the standard deviation instead of the variance). Values of σ_1 are often expressed as a percentage. Stress-I was proposed in Kruskal [8] in the case where weights are all equal to one.

A version of σ_r known as S-Stress, where disparities and distances are replaced by their squared values, has been considered to simplify computations (e.g. Takane et al. [10]). A drawback of S-Stress is a tendency to emphasize larger proximities more than smaller ones.

Once the least-squares criterion has been chosen, the minimization procedure is as follows. Starting from an initial estimate of the coordinate matrix $\mathbf{X}_0$ (e.g. the configuration of classical MDS), the computational algorithm by iteratively regressing the transformed proximities onto the distances and updating coordinate vectors in (2.8), decreases the approximation criterion until convergence to a (at least local) minimum. As an example, we can assume f in (2.5) to be a linear transformation and rewrite

$$d_{ij}(\boldsymbol{x}_i, \boldsymbol{x}_j) = \alpha\, \omega_{ij} + \beta + e'_{ij} \quad (i, j = 1, 2, \ldots, n).$$

The minimization strategy described above involves the reiteration until convergence of the following two-step process. First, starting from an initial coordinate matrix $\mathbf{X}_0$, a set of distances $d_{ij}(\boldsymbol{x}_i, \boldsymbol{x}_j)$ is computed, and the parameters α and β are estimated from a simple linear regression of the distances d_{ij} on the proximities ω_{ij}. This leads to a set of disparities calculated as

$$\hat{d}_{ij} = \hat{\alpha}\, \omega_{ij} + \hat{\beta} \quad (i, j = 1, 2, \ldots, n), \tag{2.9}$$

where $\hat{\alpha}$ and $\hat{\beta}$ are the *LS* estimates of α and β. Then, the disparities are included in the approximation criterion, and an optimization algorithm (e.g. steepest descent or majorization) is applied to find a new coordinate matrix $\mathbf{X}_1$ and a revised configuration.

An important point is picking a good initial configuration, in order to avoid the problem of *local minima* (that means the algorithm did not find the best solution). A common strategy available in many software programs is to start with several initial (random) configurations and pick the one with the lowest stress value.

For a detailed description of the approximation criteria and minimization algorithms, the reader is referred to Borg and Groenen [7]. Furthermore, a useful guide for MDS applications is Borg et al. [11]

Stability of the estimated configurations is an important indication for a global minimum. Attaining the minimum of the approximation criterion is a necessary condition for a correct choice of the dimensionality r.

Choice of r (dimensionality)

In real applications, an important question concerns the choice of the dimensionality of the spatial representation to approximate adequately the observed proximities. A complication is that the property of nested solutions does not hold for the least-squares approximation by distances. This complex issue has already been extensively addressed in Kruskal and Wish [12], who remark that it is not only a matter of attaining the minimum value for the approximation criterion (sometimes called *badness-of-fit*), since "… other considerations enter into decisions about the appropriate dimensionality, e.g. interpretability, ease of use and stability".

With respect to the numerical evaluation of the *Stress*-I values, Kruskal [8] presents a table of thresholds and he writes "… our experience with experimental and synthetic data suggests the following verbal evaluation: 20% poor, 10% fair, 5% good, 2.5% excellent, 0% perfect". Many authors (e.g. Wagenaar and Padmos [13]; Spence and Young [14]) remark that such thresholds cannot be interpreted in a mechanical manner, because stress values depend on the number of objects represented. Moreover, configurations can reveal interesting aspects of the proximity structure (e.g. nice directions of the objects in the diagram) even when the approximation function attains reasonably low values (say close to 0.2 that is 20%).

Other indices can be added to stress measures to help the choice of dimensionality (some are provided by the MDS computer programs and will be presented later in this section), and some diagrams can help the choice as well. For instance, as in principal component analysis, stress measures can be computed for several numbers of dimensions, and a diagram (scree plot) can be drawn where stress values (vertical axis) are plotted against dimensionalities (horizontal axis). The choice of the dimensionality is suggested when an evident "elbow" corresponding to r is detected in the diagram, which means that increasing dimensionality further does not decrease the stress value as much (an example will be provided in Chap. 3, Fig. 3.13).

The issues of stability and interpretation of the configurations are studied in detail in Mair et al. [15] and Borg and Mair [16], and some recently implemented utility R functions of the SMACOF package that support users to deal with these issues are presented in Mair et al. [17].

The Shepard diagram, VAF and DAF

The scatter plot with data (proximites) on the horizontal axis and model values (distances) on the vertical axis provides a useful tool for the analysis of model fitting. The disparities $\hat{d}_{ij}$ can be also added to the vertical axis in the case of optimal scaling of the proximities, and the regression line of the disparities on the proximities $\left(\hat{d}_{ij} = f(\omega_{ij})\right)$ can be drawn in the scatterplot. This plot displaying proximities, distances and disparities is known as *Shepard diagram* (an example will be provided in Chap. 3, Fig. 3.14). Outliers can be detected in the diagram, and *stress per point* can be computed by averaging all the squared residuals between any object and all the others.

The Shepard diagram is particularly informative in the case of ordinal MDS, because the transformation f is only requested to be monotone, and it is interesting to analyse the shape which it takes (e.g. roughly linear, quadratic, etc.) in scaling the data.

From least-squares regression theory, the Pearson correlation coefficient between disparities and distances $r_{\hat{d}d}$ and the raw Stress σ_r are linked by the relation

$$\sigma_r = m\, \sigma^2_{\hat{d}}\left(1 - r^2_{\hat{d}d}\right), \tag{2.10}$$

where $m = (n \times (n-1))/2$, and $\sigma^2_{\hat{d}}$ is the variance of the disparities. Hence, the higher the correlation between disparities and distances, the lower the raw stress, the better the approximation. The squared Pearson correlation coefficient $r^2_{\hat{d}d}$ is a measure of the proportion of the variance of the disparities accounted for by the distances (also indicated VAF), which is usually provided in the output of MDS computer programs jointly with the stress measures. Another measure of fit sometimes provided in the output of MDS computer programs is the dispersion of the disparities accounted for by distances (DAF), given by

$$DAF = 1 - \sigma_n. \tag{2.11}$$

Scatter plots of distances vs. disparities can be often provided: points in such a plot scatter around the bisector line, their vertical distance from the line corresponds to approximation error. When the stress measure is zero, all points lie on the bisector, and the smaller the stress, the closer the points to the bisector line. The diagram can allow to detect outliers or other anomalies. An example concerning the Thomas sociomatrix is provided in Chap. 3 (Fig. 3.15).

Whatever the approximation criterion used in the applications in this, and in Chap. 3, the quality of the configurations is evaluated by *Stress*-I, given its wide use. Other indices as VAF and DAF will be also considered.

2.3.2 *Distance Representation of Asymmetry*

When for the entries of an asymmetric proximity matrix $\mathbf{\Omega}$ we can assume that asymmetry is originating from random fluctuations, a distance model can be applied through any of the methods which disregard asymmetry recalled in Sect. 1.1.

However, the distance model cannot be applied to asymmetric proximities when asymmetry cannot be assumed due to random fluctuation, and we wish to represent both ω_{ij} and ω_{ji}. So, several approaches have been proposed to deal with asymmetry which increase the number of parameters in the reformulated distance models.

2.3.2.1 Unfolding Model

We present the unfolding model starting by a simple example concerning a comparison of maths ability between three students. Differently from the small example in Sect. 2.1, here, no variables are observed on students, but instead an experiment is performed where eighteen students in a class are asked to compare three of their classmates (s_1, s_2, s_3) in pairs and say who is the best in maths (or say whether they are equivalent).

Table 2.4 summarizes the responses obtained in the experiment.

The counts in the table represent the number of times any student in row is considered better in maths than any student in column, and they are assumed as measures of dissimilarity between students. Ties were not observed, so the corresponding counts in cells (i, j) and (j, i) sum to 18 which is the total number of students. Diagonal entries are not defined because self-evaluation is not investigated.

We are interested in analysing student differences in maths (as perceived by peers) by displaying dissimilarities as distances (instead of inner products) in a graphical representation. As for bilinear models, different sets of coordinates will be computed for rows and columns, respectively, so that each entry of the table can be directly represented and $d_{s_i s_j} \neq d_{s_j s_i}$ is allowed. Given the small number of rows and columns of Table 2.4, two dimensions (a plane) are enough for representing the dissimilarities. The coordinates of rows and columns, hereafter denoted by capital (S_1, S_2, S_3) and small (s_1, s_2, s_3) letters, respectively, are reported in Tables 2.5 and 2.6 (details on computations will be illustrated later). For instance, the dissimilarity between S_1 and s_2 (the observed frequency is 12) is equal to the distance computed by the corresponding coordinates:

Table 2.4 Maths dominance data between three students

Student	s_1	s_2	s_3
s_1	–	12	12
s_2	6	–	10
s_3	6	8	–

Table 2.5 Row coordinates for the three students

Student	First dimension	Second dimension
S_1	6.592	−0.950
S_2	−0.333	3.922
S_3	−1.686	−2.624

Table 2.6 Column coordinates for the three students

Student	First dimension	Second dimension
s_1	3.871	−0.360
s_2	−3.772	5.100
s_3	−4.672	−5.088

$$\delta_{S_1 s_2} = d_{S_1 s_2} = \sqrt{[(6.592 - (-3.772)]^2 + [(-0.950) - 5.100]^2} = 12$$

Figure 2.6a provides the distance representation of the dissimilarities in Table 2.4, where the off-diagonal dissimilarities are represented exactly by the distances. The distance of a student from himself cannot be taken into consideration, because the corresponding dissimilarity is not defined in Table 2.4.

The asymmetry of the dissimilarities is easily perceived by comparing the corresponding distances $d_{S_i s_j}$ and $d_{S_j s_i}$. For instance, student s_1 is evaluated better in maths than students s_2 and s_3 because $d_{S_1 s_2}$ and $d_{S_1 s_3}$ are larger than $d_{S_2 s_1}$ and $d_{S_3 s_1}$. Analogously, student s_2 is evaluated better in maths than student s_3 because $d_{S_2 s_3}$ is larger than $d_{S_3 s_2}$.

The relationships between dissimilarities and distances in two dimensions can be formulated more generally as follows

$$\delta_{S_i s_j} = d_{S_i s_j} = \sqrt{(x_{i1} - y_{j1})^2 + (x_{i2} - y_{j2})^2}$$

where x_{is} and y_{js} ($s = 1, 2$) are the coordinates on dimension s for student s_i in row i and student s_j in column j, respectively. If dissimilarities in the diagonal of Table 2.4 were defined, they could be depicted in the diagram; this is an important difference with respect to methods of symmetric multidimensional scaling that usually do not represent diagonal proximities.

Real applications deal with data matrices with many rows and columns, so proximities cannot be represented exactly by distances in two or three dimensions. A consequence of the approximation is that the relations between dissimilarities and distances need to be reformulated by incorporating an error component (as for bilinear models).

The *Unfolding model* for an $(n \times n)$ proximity matrix $\mathbf{\Omega}$ with *rank* q in r dimensions ($r < q$) can be formulated as

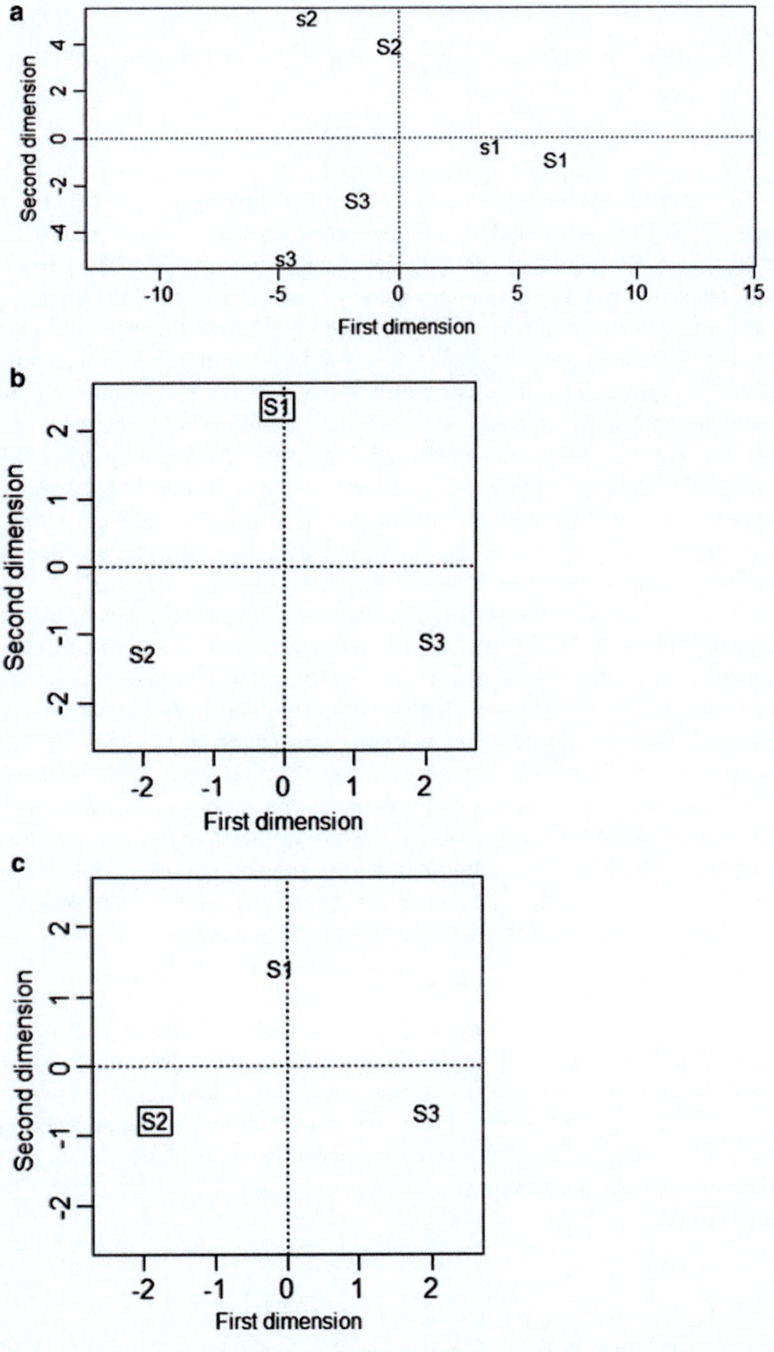

Fig. 2.6 **a** Unfolding diagram for dissimilarities in Table 2.4 represented by distances, **b** ASYMSCAL student s_1 weighted configuration for dissimilarities in Table 2.4 and **c** ASYMSCAL student s_2 weighted configuration for dissimilarities in Table 2.4

$$f(\omega_{ij}) = d^u_{ij}(\boldsymbol{x}_i, \boldsymbol{y}_j) + \varepsilon_{ij} = \sqrt{\sum_{s=1}^{r}(x_{is} - y_{is})^2 + \varepsilon_{ij}}, \quad (i,\ j\ = 1, 2, \ldots, n) \tag{2.12}$$

where f is a monotone transformation, ω_{ij} is the proximity between row i and column j, d^u_{ij} is the corresponding distance between row i and column j (and it can be $d^u_{ij} \neq d^u_{ji}$), x_{is} and y_{js} are the coordinates on dimension s of row i and column j, respectively, and ε_{ij} is an error component. In some applications, a separate transformation for each row can be considered, because the proximities can be compared only within a row (e.g. each row contains the preference ratings provided by a different subject). The distances within the two sets of row-points and column points are only implicitly defined and do not have corresponding observed proximities (for this reason, unfolding can be seen as a special case of symmetric MDS where the within-sets proximities are missing, Borg and Groenen [7, Chapter 14]). Asymmetry is derived by comparing the distances corresponding to entries ω_{ij} and ω_{ji} $(i \neq j)$. As for bilinear models, coordinate vectors $\boldsymbol{x}_i$, $\boldsymbol{y}_j$ can be collected as rows in two coordinate matrices **X** and **Y**, respectively, to be estimated.

When f is the identity transformation, the unfolding model can be viewed as the distance version of the bilinear model, where similarities are transformed into dissimilarities, and inner products are replaced by distances (the reader is referred to Heiser and de Leeuw [18] for a detailed comparison between the two models).

We remark that the unfolding model (first considered in Coombs [19, 20]) is defined more generally for rectangular matrices of preference scores (such as rank orders of preference), where rows and columns refer to persons and attitude items or stimuli, respectively. So, the proximity refers to the strength of the preference of any person for any item (or stimulus), and in the configuration, the distance of each person point (also named *ideal point*) from each item point corresponds to the preference score (the smaller the distance, the higher the preference).

Estimation of coordinate matrices **X** *and* **Y**

A good level of approximation can be obtained when the error components ε_{ij} are "small", so that the distances are as close as possible to the transformed proximities. Degenerate solutions are frequently obtained for the unfolding model when f is restricted to be a monotone function that preserves the order of the proximities. As shown for the symmetric case, the raw stress criterion (2.6) has to be modified in order to avoid degenerate solutions. The following normalization is suitable for the unfolding model

$$\text{Stress} - \text{II} = \sqrt{\frac{\sum_i \sum_j w_{ij}\left(\hat{d}_{ij} - d^u_{ij}(\boldsymbol{x}_i, \boldsymbol{y}_j)\right)^2}{\sum_i \sum_j w_{ij}\left(d^u_{ij}(\boldsymbol{x}_i, \boldsymbol{y}_j) - \bar{d}\right)^2}}, \tag{2.13}$$

where $\bar{d}$ is the average distance, so that the denominator is the deviance of the distances. *Stress*-II was proposed by Kruskal [21] in the context of factorial experiments for the case where the weights are all equal to one. A wide discussion of the criterion can be found in Kruskal and Carroll [22]. Several software programs estimate the unfolding model by *ALS* algorithms, adopting *Stress*-II as approximation criterion.

Busing et al. [23] propose a different strategy, penalizing the normalized raw stress σ_n by a multiplicative factor based on the coefficient of variation of the transformed proximities and a parameter λ. They present an algorithm for the minimization of the penalized criterion, included in IBM-SPSS software package (PREFSCAL program, for details see Sect. 2.6). The reader is referred to Busing et al. [23] and Busing [24] for a wide discussion of several criteria proposed to avoid degenerate solutions in the estimation of the unfolding model.

2.3.2.2 Models with Weighted dimensions*

In this section, two models based on weighted dimensions are briefly described: the *three-way unfolding model* and the *Asymscal model*. Some references are provided for further study, and software is suggested for their application.

Three-way unfolding model

In the unfolding model (2.12), dimensions can be differentially weighted to analyse simultaneously several proximity matrices related to the same set of objects, observed in different occasions (*three-way data*). In scalar notation, the *three-way unfolding model* can be expressed as

$$f(\omega_{ijk}) = d^{u}_{ijk}(\boldsymbol{x}_i, \boldsymbol{y}_j) + \varepsilon_{ijk} = \sqrt{\sum_{s=1}^{r} w_{ks}\left(x_{is} - y_{js}\right)^2} + \varepsilon_{ijk},$$

where ω_{ijk} is the proximity between objects i and j at occasion k, and the unknown weight w_{ks} is the nonnegative s-th diagonal element of a diagonal matrix $\mathbf{W}_k$ for each occasion k, to be estimated along with the coordinates vectors $\boldsymbol{x}_i$, $\boldsymbol{y}_j$. A diagram for the common pattern of relationships (usually named *Common Space*) is obtained by the coordinate vectors, where distances are analysed as in the two-way unfolding model (2.12). Additionally, the system of dimensional weights provided by the matrices $\mathbf{W}_k$ allows to represent the specific pattern of proximities for each occasion k, in diagrams named *individual spaces*. The distances in the individual spaces are obtained by expanding or shrinking the coordinates vectors $\boldsymbol{x}_i$, $\boldsymbol{y}_j$ by weights w_{ks} on dimension s of the configuration of the common space. Moreover, weights of different occasions can be compared in a separate diagram (*space weights*) in order to assess how much the common space can be considered a good approximation of the individual spaces. The already mentioned computer program PREFSCAL can compute the three-way unfolding model (for details on three-way algorithm see also Busing [24]). For applications of three-way unfolding to asymmetric proximity data, the reader is referred to Bove and Rocci [25], Bove [26] and Sagarra et al. [27].

Other methods to deal with three-way proximity matrices will be considered in detail in Chap. 5.

Asymscal model

In order to take into account of asymmetry, a system of weights for the dimensions can be also considered. The Asymscal model proposed by Young [28] (see also Young [29]) is a weighted Euclidean distance model, that is obtained by setting $y_{js} = x_{js}$ and adding weights v_{is} which are specific for each row and dimension in Eq. (2.12),

$$f(\omega_{ij}) = \sqrt{\sum_{s=1}^{r} v_{is}\left(x_{is} - x_{js}\right)^2} + \varepsilon_{ij} \quad (i, j = 1, \ldots, n).$$

A group configuration *(common space)* can be based on the coordinates x_{is}; moreover, for each of the n rows, a configuration of n points *(weighted space)* is obtained by shrinking or stretching the axes of the common space by the system of weights $\sqrt{v_{is}}$ $(s = 1, 2, \ldots r)$. In the weighted space of row *i*, only distances from row *i* to all the columns can be analysed. Consequently, the distance between row *i* and column *j* can be different from the distance between row *j* and column *i*, and the two distances are represented in separate spaces. For instance, for $r = 2$, the distance from row *i* to column *j* is analysed in the weighted planar configuration obtained by weights $\sqrt{v_{i1}}$, $\sqrt{v_{i2}}$, and the distance from row *j* to column *i* is analysed in the weighted planar configuration obtained by weights $\sqrt{v_{j1}}$, $\sqrt{v_{j2}}$. For the dissimilarities in Table 2.4, Fig. 2.6b and c are the weighted configurations for students s_1 and s_2, respectively. In Fig. 2.6b, only the distances from student s_1 to student s_2 and s_3 can be considered (this is the reason for the rectangle around the label s_1). In Fig. 2.6c, only the distances from student s_2 to student s_1 and s_3 can be considered. Thus, each weighted space represents the system of relationships of one object with all the others. It can be observed that the distance between student s_1 and student s_2 (equal to 4.1) in Fig. 2.6b is greater than the distance between student s_2 and student s_1 (equal to 2.8) in Fig. 2.6c. So, the only way to evaluate the asymmetry in $\mathbf{\Omega}$ by Asymscal is by comparing all the weighted configurations, which is rather cumbersome for n moderately large. Asymscal reduces to the Euclidean distance model for symmetric proximities when the weights are all equal to one.

The Asymscal model has been estimated by the function *asymscal* of the R package *Asymmetry* written by Berrie Zielman (see Sect. 2.6 for details).

Applications of Asymscal can be found in Sect. 2.5, in Young [29, Chapter 8 by Collins] and Everitt and Rabe-Hesketh [30, Chapter 6].

2.3.2.3 Slide Vector Model

One common drawback of biplot and unfolding approaches is a certain difficulty in interpreting the graphical representations of *2n* points when the number of objects becomes moderately large. Particular constrained versions of the two general models were proposed to avoid this difficulty, and some of them will be presented in Chap. 3. A special case of the unfolding model named *slide vector model*, attributed to Kruskal by De Leeuw and Heiser [31], is presented in Zielman and Heiser [32]. The model is obtained from Eq. (2.12) setting $y_{js} = x_{js} - z_s$, i.e., the configuration of the column points is constrained to be a uniform shift (or translation) of the row points in the direction of the slide vector $\mathbf{z} = (z_1, z_2, \ \ldots, \ z_r)$, while for the unfolding model, the two configurations of points are not required to be linked. Consequently, only n points and a vector are required to represent the asymmetry by the slide vector model, and such asymmetries can be analysed by the projections of the objects on the slide vector itself. This is because the model can be written in terms of squared proximities as

$$\omega_{ij}^2 = \sum_{s=1}^{r} \left(x_{is} - x_{js}\right) + \ \sum_{s=1}^{r} z_s^2 + 2\sum_{s=1}^{r} z_s\left(x_{is} - x_{js}\right) + \varepsilon_{ij},$$

where the first two terms on the right-hand side represent the symmetric component, and the third term represents the asymmetric component (such an additive decomposition will be further detailed in Chap. 3). Thus, distant points tend to be more asymmetric than points close to each other. The third term is the difference of the two inner products of vector point i and vector point j with the slide vector, respectively. This implies the *dominance* of the points with the largest projection on the slide vector (in (2.12), object i dominates object j when $f(\omega_{ij}) > f(\omega_{ji})$). Thus, the slide vector model provides a multidimensional representation of symmetry and asymmetry with common parameters (the coordinates x_{is}, x_{js}).

Figure 2.7 shows the diagram obtained by applying the slide vector model to the proximities of Table 2.4. Distances between points represent the averages $\left(\omega_{ij} + \omega_{ji}\right)/2$ of the corresponding entries ω_{ij} and ω_{ji}, all equal to 9 in Table 2.4, and the projections of the points on the axis generated by the slide vector represent the dominance relationships between students. Student s_1 has the largest projection; this means that he is more frequently judged better in maths with respect to students s_2 and s_3. Analogously, student s_2 is judged to be slightly better than student s_3 because the projection of point s_2 is slightly larger than the projection of point s_3 on the slide vector. Furthermore, the length of the slide vector, which is not too short in this illustrative example when compared with the mean distance, provides a measure of the asymmetry degree of the proximity matrix.

Since unfolding is a special case of symmetric MDS where the within-sets proximities are missing, the slide vector model can be applied by any MDS program that allows for missing values of the proximities and also for linear restrictions on the

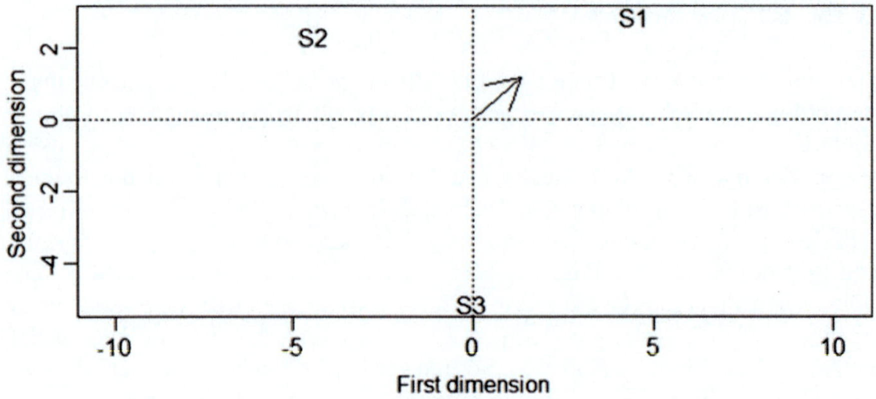

Fig. 2.7 Slide vector diagram for dissimilarities in Table 2.4

configuration. A routine named *slidevector* is available in the Asymmetry R package (for details see Sect. 2.6).

The bilinear and the distance-like approaches can be generalized to analyse simultaneously more than one data matrix related to the same set of objects (*three-way case*). Multiway extensions will be considered in more detail in Chap. 5.

2.4 External Information

Diagrams obtained by bilinear and distance-like methods can be enriched by external information possibly available for the rows of the proximity matrix. For instance, information concerning the two *external variables* maths grade in the last year and gender could be available in the example of the three students (Table 2.7).

The procedure for representing the external variables in a diagram obtained by bilinear or distance-like mehods is different according to their level of measurement. A quantitative external variable (e.g. maths grade) is linearly regressed on the coordinates of the two dimensions of the rows in the diagram (e.g. row coordinates in Table 2.5 for the unfolding diagram). The regression coefficients for the two dimensions define a vector which can be overlaid on the (bilinear or unfolding) diagram to represent the external variable. A categorical variable is represented with a point

Table 2.7 Two external variables for the three students

Student	M-grade	Gender
S_1	80	Female
S_2	60	Male
S_3	40	Male

for each category, and the coordinates of each category point are obtained as the averages of the coordinates of the row-points having the same category.

The procedure can be applied to the external variables of Table 2.7. The following linear regression model is obtained for the maths grade variable

$$\widehat{\mathrm{M}-Grade_i} = 53.047 + 4.398\ x_{i1} + 2.146\ x_{i2}$$

so that the vector of the regression coefficients of the two dimensions is (4.398, 2.146). The coordinates of the two categories of the gender variable are (6.592, –0.950) for female and (−1.009, 0.649) for male.

Figure 2.8 displays the unfolding diagram including the representation of the two external variables. Inner products of the row point vectors with the vector of the regression coefficients, augmented by the constant 53.047, represent exactly the GRADE value of the student (as a consequence of the fact that the linear regression model explains all the variance of the dependent variable M-GRADE, i.e. $R^2 = 1$). Hence, the projections of the points S_1, S_2 and S_3 on the axis generated by the vector of the regression coefficients provide the rank order of the grade levels of the three students. Student S_1 has the highest grade, S_3 has the lowest grade, and student S_2 is in between. The two categories of the external variable gender are also reported in the diagram, the female point coincides with the student row point S_1, the only female student, and the male point is right between student points S_2 and S_3 who are the two males.

The procedure for representing external variables in a diagram obtained by bilinear or distance-like methods can be also applied to the column points, and the choice seems to be mainly a subjective matter. For other approaches to take into consideration external information, we refer the reader to Borg and Groenen [7].

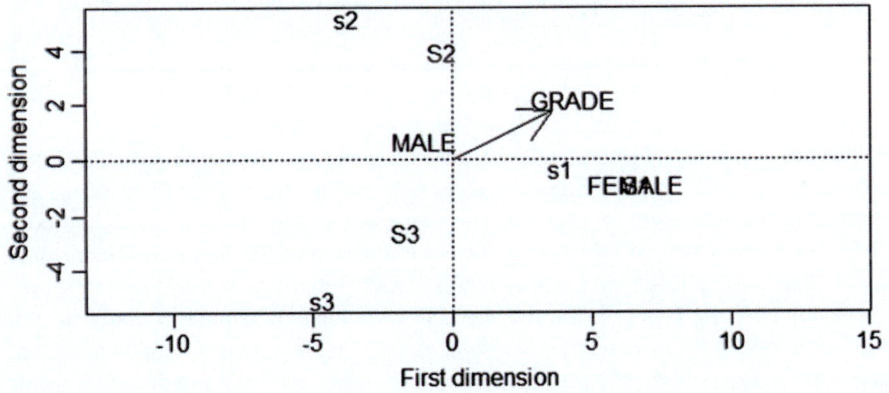

Fig. 2.8 Unfolding diagram for dissimilarities in Table 2.4 with the external variables grade and gender

2.5 Applications

In this section, we show applications of some models presented in Sects. 2.2 and 2.3 to empirical data. This allows to better understand the capabilities and limitations of the models.

2.5.1 Bilinear Model on European Languages Data

The capabilities of the bilinear model will be illustrated by using the European languages data (Eurobarometer-243 survey [33]) presented in Table 1.4. Each off-diagonal entry (i, j) is the percentage of citizens from country i who responded to speak language j "well enough to have a conversation", for sixteen European countries and their corresponding languages. The proximity can be assumed as a measure of closeness (or similarity). Diagonal entries are conventionally set to 100. The aim of the analysis is to provide a graphical representation of the relationships between countries/languages (a more detailed analysis of these data will be presented in Sect. 4.5.2.).

Data reflect a strongly asymmetric situation, and the most remarkable feature is the presence of 173 off-diagonal cells (72.1%) with percentages equal to zero. Most of the nonzero percentages are located in the columns corresponding to the three most spoken European languages, English, German and French, respectively. English is the most widely foreign language spoken on average, followed by German and French, although to a very much lesser extent. The largest percentages in the column of the spoken English language correspond to the rows of Sweden, Denmark and The Netherlands, while the smallest percentages correspond to Hungary and Bulgaria. In the column of the German language, The Netherlands, Hungary, Czech Republic and Slovakia display the highest percentages, while Spain, Italy, Poland and Romania the lowest. For the French language, the highest percentages are registered for United Kingdom, Romania, Portugal, Italy and Spain, while the lowest percentages are observed for people from the Scandinavian countries and Hungary.

The bilinear model was estimated by the R routine presented in Sect. 2.6, adapted to the data in Table 1.4. The singular values obtained for the sixteen dimensions are reported in the *scree plot* of Fig. 2.9. The shape of the line shows an evident elbow at dimension 2, which suggests to analyse the diagram of the first two dimensions, reported in Fig. 2.10. Capital letters identify rows (countries); small letters represent columns (languages) in the diagram. The variance explained by the first two dimensions (VAF) is low (32.6%) but some interesting information can be detected. In particular, the ranking of the percentages in the columns corresponding to English and German spoken as foreign languages can be detected with a good approximation by the projection of the country points (capital letters) on the two vectors (reported in the diagram) determined by the origin and the language points (small letters) uk and de, corresponding to English and German, respectively.

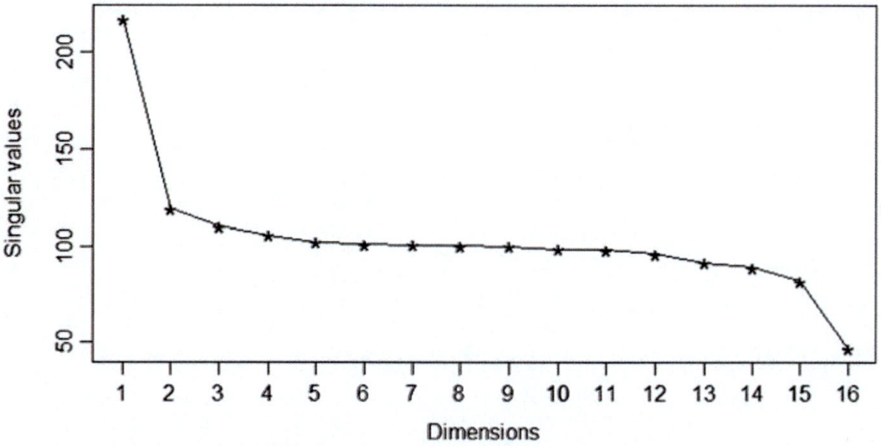

Fig. 2.9 Scree plot of the bilinear model for the European languages data

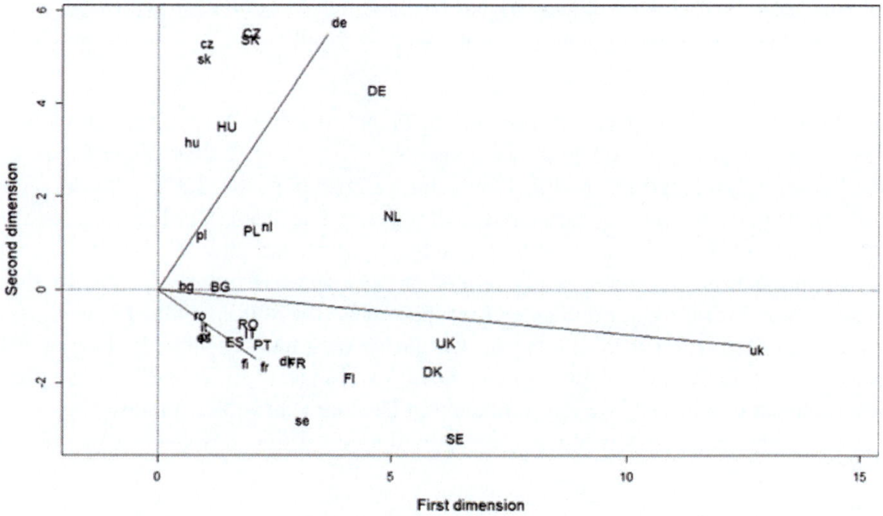

Fig. 2.10 Representation of the European languages data by the bilinear model (first and second dimension, capital letters for countries, small letters for languages)

For instance, for the English language, Sweden (SE), Denmark (DK) and The Netherlands (NL) have the longest projections on the vector from the origin to uk (in addition to United Kingdom), while Hungary and Bulgaria are the smallest. Analogously, the ranking of the percentages in the column of the German language can be detected by the projections of the country points on the vector from the origin to the language point de. Moreover, the large inner products of the Czech Republic and Slovakia in the diagram reflect the corresponding percentages in Table 1.4. The

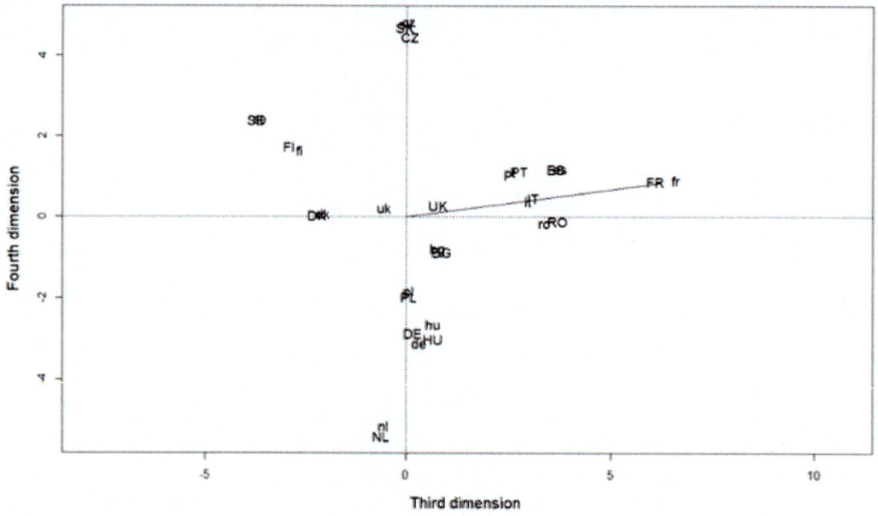

Fig. 2.11 Representation of European languages data by the bilinear model (third and fourth dimension, capital letters for countries, small letters for languages)

ranking of the percentages corresponding to French is not well represented, so the diagram of the third and fourth dimensions displayed in Fig. 2.11 is also analysed. The variance explained by the third and fourth dimensions is 12.3%. This second diagram trivially reflects the percentages along the main diagonal of Table 1.4 (all equal to 100%) because the country and language points are almost coincident.

The projections of the country points on the vector from the origin to the language point fr reproduce quite well (except for UK) the ranking of the percentages corresponding to the French language. Countries with native Romance languages (Romania, Italy, Portugal and Spain) have the longest projections, while the Scandinavian countries (Sweden, Finland and Denmark) have the smallest.

Other interesting aspects of this data set will be investigated in Sect. 4.5.2.

2.5.2 Distance Representations of Managers Data

The capabilities of some models considered in Sect. 2.3 will be illustrated by using managers data by Krackhardt [34] presented in Chap. 1 (Table 1.2). Each off-diagonal entry (i, j) is the number of managers who responded that manager i goes to manager j for help and advice at work. The proximity is assumed as a measure of closeness (or similarity) between managers. Aim of this application is to analyse the relationships for help and advice between managers at work, and the centrality role of the president and vice-presidents. For the application of cluster analysis methods to these data, the reader can refer to Sect. 4.5.1.

As a first step, it is interesting to analyse and compare the row and column totals of Table 1.2, because they provide a preliminary indication of the tendency of managers to ask or to be asked for help and advice. The analysis can be done looking at the diagram in Fig. 2.12, where each manager is identified by its row label in Table 1.2. Coordinates of the points are the corresponding row totals (number of times manager asks for help according to the opinions of the other 20 managers) and column totals (number of times manager is asked for help), on the horizontal and vertical axes, respectively. The bisector line (equation $y = x$) is also drawn in the plane to allow for a comparison between totals. Manager points close to the line have similar row and column totals, while manager points above (below) the line have the column (row) totals higher than row (column) totals.

Thus, for instance, manager 7 (the president of the firm) and managers 2, 14, 18 and 21 (the four vice-presidents) are above the bisector line, which means that the number of requests for help and advice they receive is higher than the number of requests they make, according to the opinions of the other managers. This result is not surprising given the roles of the five managers in the firm. Managers 2 and 18 have the highest number of requests for help and advice, despite their two departments have only three and two supervisors, respectively. It follows that managers 2 and 18 receive requests also from supervisors of other departments. Moreover, it is interesting to notice that supervisor 11 is positioned above the bisector line, he belongs to the small department of the vice-president 18 but receives many requests for help from managers of other departments.

The pattern of help relationships between the twenty-one managers can be analysed in detail by the unfolding model presented in Sect. 2.3.2.1. A proximity matrix

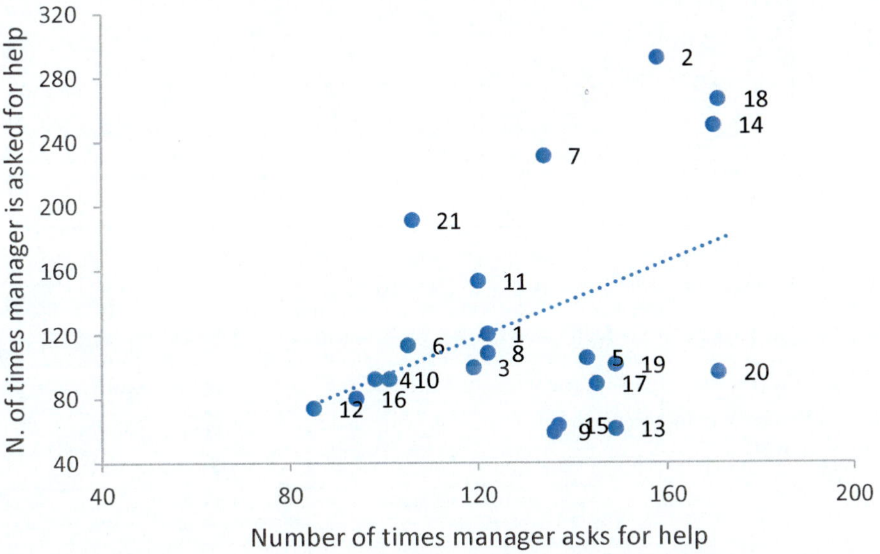

Fig. 2.12 Scatter diagram of row and columns totals of managers data in Table 1.2

measuring dissimilarities between managers was computed by subtracting any off-diagonal entry of the matrix in Table 1.2 from the maximum value 21, so that the entry (i, j) measures how many managers responded that manager i *does not go* to manager j for help or advice at work. Diagonal entries are missing, so the corresponding weights in the approximation criterion have been set equal to zero. The coordinates for the configurations representing the proximities have been computed by the PREFSCAL routine in the IBM-SPSS package with default options.

Metric unfolding (model (2.8) when f is the identity transformation) has been fitted by varying the number of dimensions. The scree plot is shown in Fig. 2.13. To avoid local minima, the stability of the solutions has been assessed starting from the classical scaling and several multiple random initial configurations.

On the basis of both the "elbow" criterion in the scree plot and the level of the Stress-I values, the solutions for two and three dimensions have been selected to be analysed. The Stress-I, $r_{\hat{d}d}$, VAF and DAF values are provided in Table 2.8.

The Stress-I value in two dimensions is not high but acceptable; the values of the other indices are quite satisfactory. The errors of approximation can be analysed in Fig. 2.14. Residuals of small dissimilarities are in some cases high, whereas medium and high dissimilarities are usually quite well represented, evident outliers are missing. A detailed analysis of the stress per point can be conducted in the output (not reported here) of the PREFSCAL routine, based on the 21 $\times$ 21 table of the

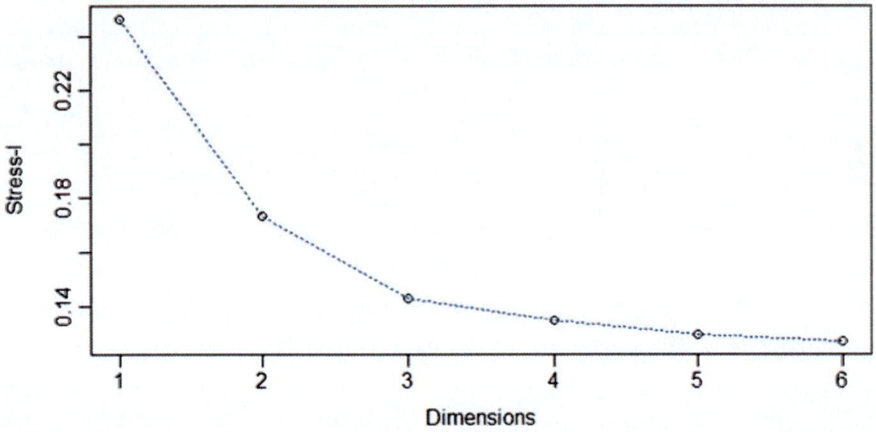

Fig. 2.13 Scree plot of the unfolding model for several dimensionalities (managers data)

Table 2.8 Statistics for the unfolding model

	Two dimensions	Three dimensions
Stress-I	0.173	0.143
$r_{\hat{d}d}$	0.880	0.926
VAF	0.774	0.857
DAF	0.971	0.980

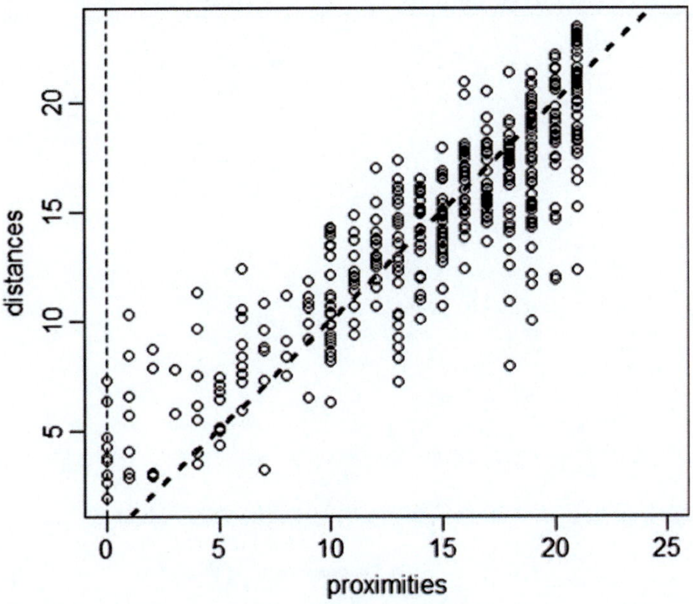

Fig. 2.14 Residual plot for the unfolding model in two dimensions for managers data

squared residuals. High residuals for small proximities regard almost exclusively entries from managers to the president and to the vice-presidents. In particular, the columns corresponding to the president and the vice-presidents 14 and 21 present in many cases high residuals, so their positions in the configurations need to be analysed carefully.

The diagram for two dimensions is reported in Fig. 2.15, points labelled by numbers represent the rows and points labelled by the letter *m* plus numbers represent columns. First, we remark that points identifying supervisors in columns are located far from the centre of the configuration, whereas most of the row-points are closer to the centre. This means that supervisors are not frequently approached for help and advice. Moreover, some supervisors appear to be quite isolated because they do not ask many people for help (e.g. row point of manager 12 is also far from the central area of the configuration). Row and column points corresponding to the president and to the three vice-presidents 2, 14 and 18 are positioned in the centre of the diagram which means that they are the main destinations of the requests for advice, while vice-president 21 is located in the direction of the fourth quadrant. The analysis of the positions of these five column points allows us to detect a sort of hierarchy, where vice-president 2 is approached for advice more frequently than the other vice-presidents (as already observed in Fig. 2.12), and, in turn, he appears more linked to the president than the others. Furthermore, vice-presidents 14 and 18 appear more central than vice-president 21. Note that a similar result is reported in Krackhardt [34, p. 119].

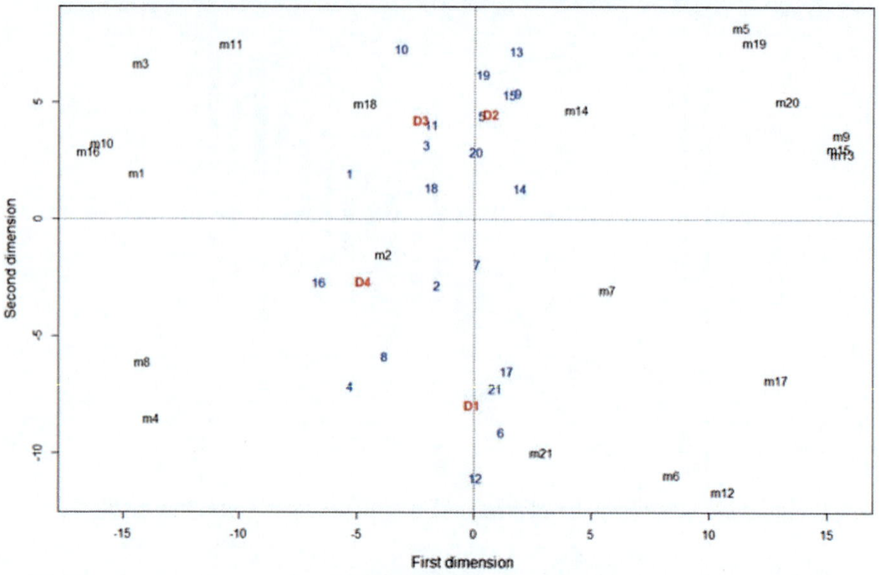

Fig. 2.15 Representation of managers data in Table 1.2 by the unfolding model in two dimensions (blue labels for the rows, black labels for the columns, red labels for departments)

The external categorical variable regarding the department the managers belong to is also represented in the diagram by the procedure described in Sect. 2.4 (labels D1, D2, D3, D4). The distance between managers reflects quite well the departments they belong to, especially for the largest ones (Department 1 in the fourth quadrant and Department 2 in the first quadrant), with just a few exceptions (managers 3 and 8). The position of the vice-president 21, close to the managers of Department 1 that he heads up, suggests that his advice is almost exclusively oriented to the managers of his Department (a similar result was obtained by Okada [35]).

The approximation of the data improves quite a bit for the three dimension solution, as shown in Table 2.8. Figure 2.16 shows that the improvement mainly concerns the medium-large dissimilarites. The configuration for three dimensions is presented in Fig. 2.17. A segment perpendicular to the plane $z = 0$ locates each manager point to facilitate visualization. In spite of some difficulties in the inspection of the diagram, most of the spatial relationships seen in two dimensions are confirmed. The centrality of vice-presidents 2, 18, and to a lesser extent, 14 is confirmed and can be detected by the positions of their corresponding points in the central space. A difference emerges with respect to the president position, who is closer to vice-president 21 than to vice-president 2 (this is confirmed by the entries in the proximity matrix).

The relationships among the president and the vice-presidents can be also analysed by the Asymscal model. The model has been fitted by using the function *asymscal* of the R package *Asymmetry*. The measures of fitting for the solutions in two and three dimensions are provided in Table 2.9.

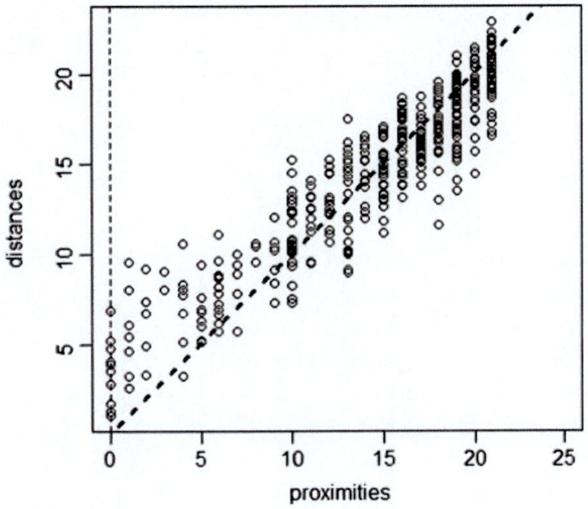

Fig. 2.16 Residual plot for the unfolding model in three dimensions for managers data

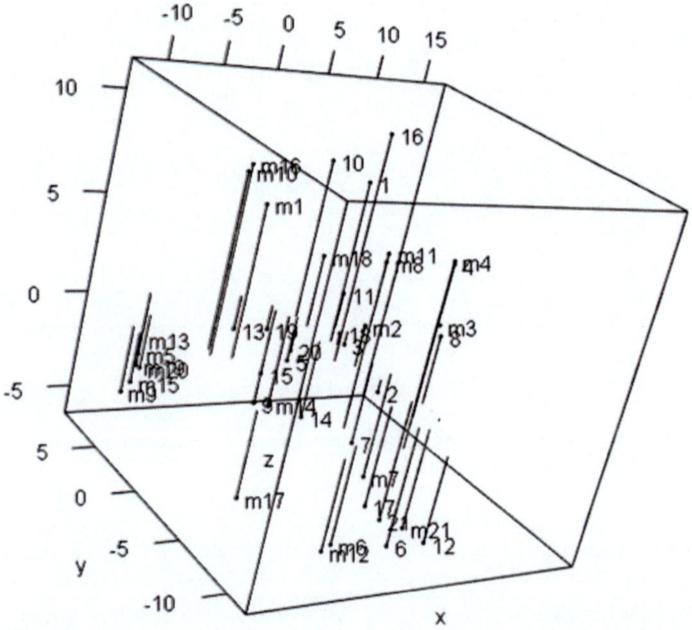

Fig. 2.17 Representation of managers data in Table 1.2 by the unfolding model in three dimensions

Figure 2.18 shows the weighted space of the president in two dimensions, where only distances from point 7 (framed in a rectangle) to all the other points can be analysed (the distances approximate proximities in row 7 of Table 1.2). Lines from the president (7) to the vice-presidents are drawn to visualize the distances. Moreover,

Table 2.9 Statistics for the Asymscal model

	Two dimensions	Three dimensions
Stress-I	0.263	0.176
$r_{\hat{d}d}$	0.833	0.910
VAF	0.694	0.829
DAF	0.935	0.970

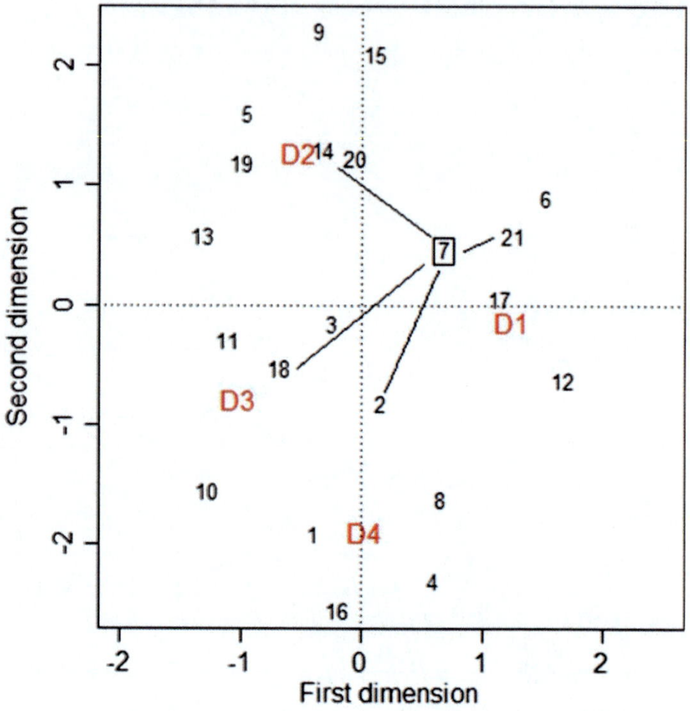

Fig. 2.18 Asymscal president weighted configuration in two dimensions for proximities in row 7 of Table 1.2 (red labels for Departments)

departments are represented as red labels with coordinates computed as averages of the coordinates of the managers belonging to the departments themselves.

In spite of the high stress value for the Asymscal solution in two dimensions, the proximities between the president and vice-presidents are satisfactorily represented. President goes to vice-president 21 for help and advice more frequently than to other vice-presidents, especially vice-president 18.

Another aspect pointed out by the solution of the Asymscal model in two dimensions is the position of vice-president 21 with respect to the supervisors. Figure 2.19 shows the weighted space of the vice-president 21 where the analysis of the distances confirms that the advices of vice-president 21 are almost exclusively given to the

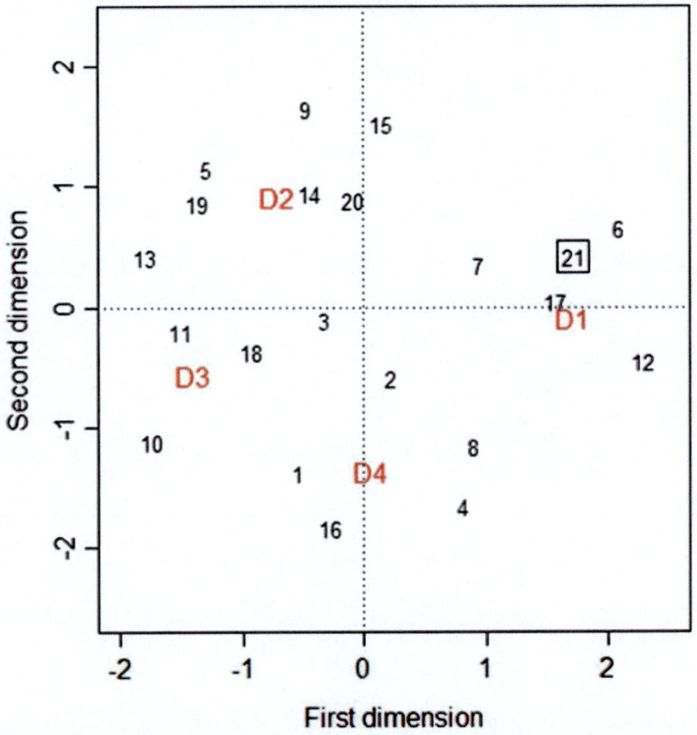

Fig. 2.19 Asymscal vice-president 21 weighted configuration in two dimensions for proximities in row 21 of Table 1.2 (red labels for Departments)

supervisors of his Department (labels 6, 12, 17 and to a lesser extent 8), in addition to the president (7).

The weighted space configurations of the president and the vice-president 21 obtained from the Asymscal model in three dimensions (not shown here) confirm the results in two dimensions. The weights to compute the two weighted spaces from the group configuration result $(0.0, 34.78, 19.90)$ and $(31.01, 24.17, 0.0)$, respectively, so that distances are analysed in two planar configurations of the three-dimensional space.

The asymmetry analysis performed by the unfolding model and the Asymscal model is not immediate as it involves the comparison of the distances between many pairs of points. The slide vector model in some cases can allow to avoid this difficulty; for this reason, it is applied to the managers data. The estimation of the model is performed by the function *slidevector* of the R package *Asymmetry*. The fit values in two dimensions are Stress-I $= 0.30$, $r_{\hat{d}d} = 0.77$, VAF $= 0.59$ and DAF $= 0.92$, and Fig. 2.20 shows the configuration obtained. The distances represent the departments the managers belong to quite well; however, the projections of the points on the slide vector rarely allow to detect the asymmetry appropriately. The fit values in three

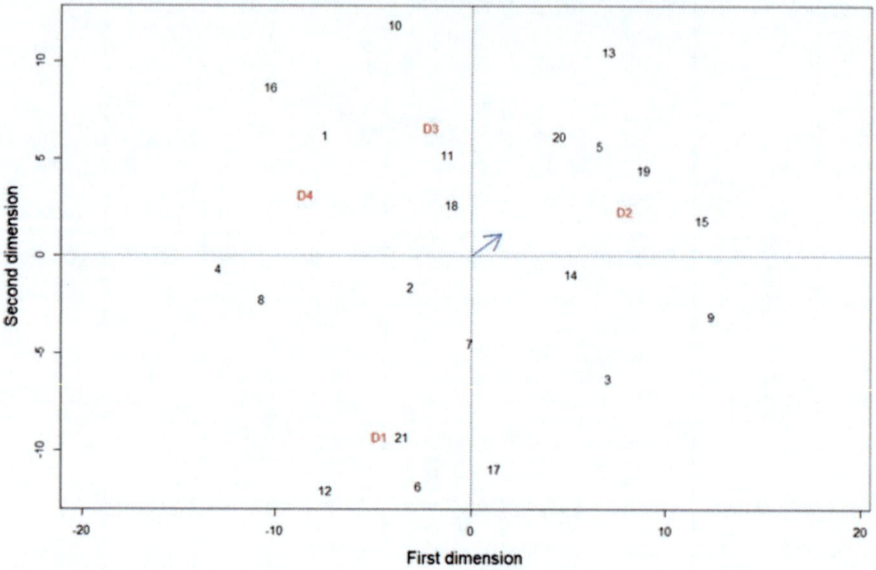

Fig. 2.20 Representation of managers data in Table 1.2 by the slide vector model in two dimensions (red points for Departments)

dimensions improve (Stress-I = 0.22, $r_{\hat{d}d} = 0.85$, VAF = 0.73, DAF = 0.95) but the analysis of the asymmetry is very difficult. These results show that the slide vector model does not allow to simplify the graphical interpretation of the asymmetry with respect to the unfolding and the Asymscal models for the managers data.

For detailed analyses of this data set, the reader is referred to Krackhardt [34], Okada [35] and Wasserman and Faust [36].

2.6 Software

In this section, informations concerning computer programs which allow for the application of some models and methods presented in Chap. 2 are provided. The R scripts serve to allow an easy application of the methods, and they might not be the most efficient way for model estimations.

2.6.1 Bilinear Methods

2.6.1.1 SVD and Bilinear Model

The bilinear model (2.3) can be applied by any computer program including routines for SVD computation and graphical representation.

For instance, the computation of the diagram in Fig. 2.5a was obtained by using the object-oriented programming language R, which can be freely downloaded from the R-project website (http://www.r-project.org). In order to work efficiently with R, an appropriate editor is required (e.g. RStudio; see http://rstudio.org and Verzani [37]). The following R commands reproduce the diagram.

```
# Data input
#
omega <- matrix(c(36,33,39,12,26,14,7,5,8),3,3,byrow = TRUE)
#
# Computation of matrices X (rowcoord) and Y (colcoord) by
symmetric
# assignment and 90° clockwise rotation
#
rotation<- matrix(c(0,1,0,-1,0,0,0,0,1),3,3,byrow = TRUE)
rowcoord <- (svd(omega)$u%*%diag(sqrt(svd(omega)$d)))%*%t(rotation)
colcoord <- (svd(omega)$v%*%diag(sqrt(svd(omega)$d)))%*%t(rotation)
#
require(ggplot2)
require(ggthemes)
#
# Computation of the diagram in two dimensions
#
#
dat=data.frame(rbind(rowcoord,colcoord))
rownames(dat) <- c("A","B","C","a","b","c")
#
theme_set(theme_bw())
ggplot(dat, aes(dat[,1],dat[,2], labels(rownames(dat))))+
  geom_point()+
  geom_text(aes(label=rownames(dat)),hjust=-0.3, vjust=-0.3)+
  scale_x_continuous(limits=c(-4,2), breaks=c(-4:3)) +
  scale_y_continuous(limits=c(-0.05,8), breaks=c(0:9))+
  coord_fixed(ratio=1)+
  geom_hline(yintercept=0, linetype="dotted")+
  geom_vline(xintercept=0, linetype="dotted")+
  geom_segment(aes(x=0,y=0, xend=dat[1,1],yend=dat[1,2]))+
  geom_segment(aes(x=0,y=0, xend=dat[2,1],yend=dat[2,2]))+
  geom_segment(aes(x=0,y=0, xend=dat[3,1],yend=dat[3,2]))+
  labs(x="First dimension", y="Second dimension")+
  theme(panel.grid.minor = element_line(colour="White"))+
  theme(panel.grid.major = element_line(colour="White"))+
  theme(axis.title=element_text(size=13))
```

Researchers working with the R language will find very useful the software information concerning biplot models in Greenacre [3] and Gower et al. [1]. Both books are provided with a supporting website containing additional material and the script files of the R code to reproduce the figures described in the books.

Moreover, the program MULTBIPLOT is available for free download at the website http://biplot.dep.usal.es/contactar.html (Department of Statistics, Salamanca University, Spain). The program allows to perform several types of biplot analyses (e.g. classical biplot, simple and multiple correspondence analysis, Canonical/MANOVA biplot, logistic biplots for binary data, etc.).

2.6.1.2 Okada–Tsurumi Model

The model can be applied by any computer program including routines for the SVD computation and graphical representation. The following R commands reproduce the diagrams in Fig. 2.5b and c

```
# Data input
#
#
omega <- matrix(c(36,33,39,16,26,14,7,5,8),3,3,byrow = TRUE)
#
# Computation of matrices X (rowcoord) and Y (colcoord) by
symmetric
# assignment
#
rowcoord <- (svd(omega)$u%*%diag(sqrt(svd(omega)$d)))
colcoord <- (svd(omega)$v%*%diag(sqrt(svd(omega)$d)))
#
require(ggplot2)
require(ggthemes)
#
# Computation of the diagrams for two dimensions
#
#
dat=data.frame(cbind(rowcoord,colcoord))
rownames(dat) <- c("A","B","C")
#
theme_set(theme_bw())
ggplot(dat, aes(dat[,1],dat[,4], labels(rownames(dat))))+
  geom_point()+
  geom_text(aes(label=rownames(dat)),hjust=-0.3, vjust=-0.3)+
  scale_x_continuous(limits=c(-8,2), breaks=c(-8:2))+
  scale_y_continuous(limits=c(-6,2), breaks=c(-6:2))+
  coord_fixed(ratio=1)+
  geom_hline(yintercept=0, linetype="dotted")+
  geom_vline(xintercept=0, linetype="dotted")+
  geom_segment(aes(x=dat[1,1],y=0, xend=dat[1,1],yend=dat[3,4]),
    linetype="dotdash")+
  geom_segment(aes(x=0,y=dat[3,4], xend=dat[1,1],yend=dat[3,4]),
    linetype="dotdash")+
```

```
    geom_segment(aes(x=dat[3,1],y=0, xend=dat[3,1],yend=dat[1,4]),
      linetype="longdash")+
    geom_segment(aes(x=0,y=dat[1,4], xend=dat[3,1],yend=dat[1,4]),
      linetype="longdash")+
    labs(x="First dimension - g1", y="First dimension - v1")+
    theme(panel.grid.minor = element_line(colour="White"))+
    theme(panel.grid.major = element_line(colour="White"))+
    theme(axis.title=element_text(size=13))
#
#
theme_set(theme_bw())
ggplot(dat, aes(dat[,2],dat[,5], labels(rownames(dat))))+
    geom_point()+
    geom_text(aes(label=rownames(dat)),hjust=-0.3, vjust=-0.3)+
    scale_x_continuous(limits=c(-8,2), breaks=c(-8:2))+
    scale_y_continuous(limits=c(-6,2), breaks=c(-6:2))+
    coord_fixed(ratio=1)+
    geom_hline(yintercept=0, linetype="dotted")+
    geom_vline(xintercept=0, linetype="dotted")+
    geom_segment(aes(x=dat[1,2],y=0, xend=dat[1,2],yend=dat[3,5]),
      linetype="dotdash")+
    geom_segment(aes(x=0,y=dat[3,5], xend=dat[1,2],yend=dat[3,5]),
      linetype="dotdash")+
    geom_segment(aes(x=dat[3,2],y=0, xend=dat[3,2],yend=dat[1,5]),
      linetype="longdash")+
    geom_segment(aes(x=0,y=dat[1,5], xend=dat[3,2],yend=dat[1,5]),
      linetype="longdash")+
    labs(x="Second dimension - g2", y="Second dimension - v2")+
    theme(panel.grid.minor = element_line(colour="White"))+
    theme(panel.grid.major = element_line(colour="White"))+
    theme(axis.title=element_text(size=13))
```

2.6.2 *Distance Representation of Asymmetry*

2.6.2.1 Unfolding Model

Several (standalone or package) computer programs doing MDS include routines for computation of the two-way and three-way multidimensional unfolding models (the reader is referred to Borg & Groenen [7], Appendix A). In the application to managers data in Sect. 2.5.2, the estimation of the unfolding model was performed by the procedure named PREFSCAL, included in the add-on module CATEGORIES of the IBM-SPSS [38] statistical package. The PREFSCAL algorithm minimizes a penalized stress measure proposed in Busing et al. [23] that helps to avoid degenerate solutions. Another advantage is that PREFSCAL allows you to put restrictions on the coordinate matrices **X** and **Y**. Several types of statistics and plots are provided (e.g. stress measures and other goodness-of-fit measures like VAF and DAF, stress decompositions, coordinate matrices, distances, transformed proximities, stress plots, common space scatterplots, individual space weight scatterplots, individual spaces scatterplots, transformation plots and Shepard residual plots). Missing values can be treated by defining a matrix of weights for the proximities. The following commands produced the output for the solution of the unfolding model in two dimensions (default options) in Sect. 2.5.2.

```
DATASET ACTIVATE Dataset1.
PREFSCAL VARIABLES=m1 m2 m3 m4 m5 m6 m7 m8 m9 m10 m11 m12 m13 m14 m15
m16 m17 m18 m19 m20 m21
  /WEIGHTS=w1 w2 w3 w4 w5 w6 w7 w8 w9 w10 w11 w12 w13 w14 w15 w16 w17
w18 w19 w20 w21
  /INITIAL=CLASSICAL(TRIANGLE)
  /CONDITION=MATRIX
  /TRANSFORMATION=NONE
  /PROXIMITIES=DISSIMILARITIES
  /CRITERIA=DIMENSIONS(2,2) DIFFSTRESS(.000001) MINSTRESS(.00001)
MAXITER(50000)
  /PENALTY=LAMBDA(0.5) OMEGA(1.0)
  /PRINT=INPUT INITIAL MEASURES DECOMPOSITION TRANSFORMATIONS
COMMON DISTANCES
  /PLOT=COMMON TRANSFORMATIONS(ALL) SHEPARD(ALL) FIT(ALL) RESID-
UALS(ALL).
```

Researchers working with the R language can estimate the unfolding models by three functions of the Smacof package (*unfolding, prefscal, smacofRect*). In particular, the *unfolding* function is one of the recent major updates of the package [17], which provides many of the outputs considered in the application to the managers data in Sect. 2.5. The following R commands can be used to obtain a diagram very similar to Fig. 2.15.

```
library(smacof)
#
# Data input of the matrix omega=(21 - proximities in Table 1.2)
# from clipboard (diagonal entries set to 0) and weight matrix defi-
nition
#
#
omega <- read.table("clipboard")
weight <- (diag(21) == 0)
#
# Unfolding model estimation and plot in two dimensions
#
unf_omega <- unfolding(omega, type="ratio", weightmat = weight)
plot(unf_omega)
```

2.6.2.2 Asymscal

The estimation of the Asymscal model can be done by the function *asymscal* of the R package Asymmetry written by Berrie Zielman, described at https://rdrr.io/cran/asymmetry/. The package provides the slide vector model, a scaling model with unique dimensions and the asymscal model for asymmetric multidimensional scaling. Furthermore, a heat map for skew-symmetric data and the decomposition of asymmetry are provided for the analysis of asymmetric tables.

The following R commands can be used to obtain the diagrams in Figs. 2.6b and c (except for the two frames around labels).

```
library(asymmetry)
#
# Data input
#
omega <- matrix(c(0,12,12,6,0,10,6,8,0),3,3,byrow = TRUE)
#
# Labels definitions
#
rownames(omega) <- c("S1","S2","S3")
#
# Asymscal model estimation
#
t<-asymscal(omega, ndim = 2, itmax = 10000, eps = 1e-10)
#
# Outputs: group configuration coordinates and weights
t$gspace
t$cweights
round(t$cweights, 3)
#
# Plot of group configuration
#
plot(t, plot.type = "confplot")
#
# Individual spaces computations
#
# First individual space
#
t1 <- matrix(c(8.422557,0,0,8.506004),2,2,byrow = TRUE)
config=t$gspace%*%sqrt(t1)
plot(config, type="n", xlab = "First dimension",ylab = "Second
dimension",
    xlim=c(-2.5,2.5), ylim=c(-2.5,2.5))
abline(h=0, v=0, lty="dotted")
text(config, cex=0.8, labels=rownames(omega))
#
# Second individual space
#
t2 <- matrix(c(7.075449,0,0,3.018532),2,2,byrow = TRUE)
config=t$gspace%*%sqrt(t2)
plot(config, type="n", xlab = "First dimension",ylab = "Second
dimension",
    xlim=c(-2.5,2.5), ylim=c(-2.5,2.5))
abline(h=0, v=0, lty="dotted")
text(config, cex=0.8, labels=rownames(omega))
#
```

The estimation of the Asymscal model can be done by choosing the options SHAPE = asymmetric and MODEL = Ascal in the procedure named ALSCAL, included in the IBM-SPSS Base [38] statistical package (version 26). The procedure can be applied to proximity matrices with $n > 3$ elements. ALSCAL differs from other MDS programs in minimizing S-Stress rather than Stress, thereby fitting squared distances to squared dissimilarities.

2.6.2.3 Slide Vector Model

The slide vector model can be fitted using the function *slidevector* of the R package Asymmetry written by Berrie Zielman, described at https://rdrr.io/cran/asymmetry/ (see previous section for a brief description of the main features of the Asymmetry package).

The following R commands can be used to obtain the diagram in Fig. 2.7 (axes orientation can differ).

```
library(asymmetry)
#
# Data input
#
omega <- matrix(c(0,12,12,6,0,10,6,8,0),3,3,byrow = TRUE)
#
# Labels definitions
#
rownames(omega) <- c("S1","S2","S3")
#
# Slidevector model estimation
#
slv <- slidevector(omega, ndim = 2)
#
# Outputs: configuration and slide-vector coordinatesn
#
summary(slv)
#
# Plot of the configuration
#
plot(slv, type="n", asp=1, xlab = "First dimension",ylab = "Second
dimension")
abline(h=0, v=0, lty="dotted")
```

References

1. Gower, J. C., Lubbe, S., & Le Roux, N. (2011). *Understanding biplots*. Wiley.
2. Gabriel, K. R. (1971). The biplot graphic display of matrices with application to principal component analysis. *Biometrika, 58*(3), 453–467.
3. Greenacre, M. J. (2010). *Biplots in practice*. BBVA Foundation.

4. Okada, A., & Tsurumi, H. (2012). Asymmetric multidimensional scaling of brand switching among margarine brands. *Behaviormetrika, 39*, 111–126.
5. Torgerson, W. S. (1952). Multidimensional scaling: I theory and method. *Psychometrika, 17*, 401–419.
6. Gower, J. C. (1966). Some distance properties of latent root and vector methods used in multivariate analysis. *Biometrika, 53*, 325–338.
7. Borg, I., & Groenen, P.J. F. (2005). *Modern multidimensional scaling. Theory and applications.* (2nd ed.). Springer.
8. Kruskal, J. B. (1964). Multidimensional scaling by optimizing goodness of fit to a nonmetric hypothesis. *Psychometrika, 29*, 1–29.
9. Sammon, J. W. (1969). A nonlinear mapping for data structure analysis. *IEEE Transactions on Computers*, C-18 (5), 401–409.
10. Takane, Y., Young, F. W., & de Leeuw, J. (1977). Nonmetric individual differences multidimensional scaling: An alternating least squares method with optimal scaling features. *Psychometrika, 42*, 7–67.
11. Borg, I., Groenen, P. J. F., & Mair P. (2018) *Applied Multidimensional Scaling and Unfolding* (2nd edn.). Springer.
12. Kruskal, J. B., & Wish, M. (1978). *Multidimensional scaling*. Sage.
13. Wagenaar, W. A., & Padmos, P. (1971). Quantitative interpretation of stages in Kruskal's multidimensional scaling technique. *British Journal of Mathematical and Statistical Psychology, 24*, 101–110.
14. Spence, I., & Young, F. W. (1978). Monte Carlo studies in nonmetric scaling. *Psychometrika, 43*, 115–117.
15. Mair, P., Borg, I., & Rusch, T. (2016). Goodness-of-fit assessment in multidimensional scaling and unfolding. *Multivariate Behavioral Research, 51*, 772–789.
16. Borg, I., & Mair, P. (2017). The choice of initial configurations in multidimensional scaling: Local minima, fit, and interpretability. *Austrian Journal of Statistics, 46*, 19–32.
17. Mair, P., Groenen, P. J. F., & De Leeuw, J. (2020). *More on multidimensional scaling and unfolding in R: smacof version 2.* https://cran.r-project.org/web/packages/smacof/vignettes/smacof.pdf
18. Heiser, W. J., & de Leeuw, J. (1981). Multidimensional mapping of preference data. *Mathématique et Sciences Humaines, 19*(73), 39–96.
19. Coombs, C. H. (1950). Psychological scaling without a unit of measurement. *Psychological Review, 57*, 148–158.
20. Coombs, C. H. (1964). *A theory of data.* Wiley.
21. Kruskal, J. B. (1965). Analysis of factorial experiments by estimating monotone transformations of the data. *Journal of the Royal Statistical Society Series B (methodological), 27*, 251–263.
22. Kruskal, J. B., & Carroll, J. D. (1969). Geometrical models and badness-of-fit functions. In P. R. Krishnaiah (Ed.), *Multivariate analysis* (Vol. 2, pp. 639–671). Academic Press.
23. Busing, F. M. T. A., Groenen, P. J. F., & Heiser, W. J. (2005). Avoiding degeneracy in multidimensional unfolding by penalizing on the coefficient of variation. *Psychometrika, 70*, 71–98.
24. Busing, F. M. T. A. (2010). *Advances in multidimensional unfolding* (PhD Thesis). Faculty of Social and Behavioural Sciences, Leiden University
25. Bove, G., & Rocci, R. (1999). Methods for asymmetric three-way scaling. In M. Vichi & O. Opitz (Eds.), *Classification and data analysis. Theory and application* (pp. 131–138). Springer.
26. Bove, G. (2014). Asymmetries in organizational structures. In D. Vicari, A. Okada, G. Ragozini, & C. Weihs (Eds.), *Analysis and modeling of complex data in behavioral and social sciences* (pp. 65–72). Springer-Verlag.
27. Sagarra, M., Busing, F. M. T. A., Mar-Molinero, C., & Rialp, J. (2018). Assessing the asymmetric effects on branch rivalry of Spanish financial sector restructuring. *Advances in Data Analysis and Classification, 12*, 131–153.

28. Young, F. W. (1975). An asymmetric Euclidean model for multi-process asymmetric data. In *Proceedings of the US-Japan Seminar on the Theory, Methods and Applications of Multidimensional Scaling and Related Techniques* (pp. 79–88). University of California.
29. Young, F. W. (1987). *Multidimensional scaling: History, theory, and applications.* R. M. Hamer (Ed.). Lawrence Erlbaum Associates Inc.
30. Everitt, B. S., & Rabe-Hesketh, S. (1997). *The analysis of proximity data.* Arnold.
31. De Leeuw, J., & Heiser, W. J. (1982). Theory of multidimensional scaling. In P. R. Krishnaiah & L. N. Kanal (Eds.), *Handbook of statistics* (Vol. 2, pp. 285–316). North Holland.
32. Zielman, B., & Heiser, W. J. (1993). Analysis of asymmetry by a slide-vector. *Psychometrika, 58*, 101–114.
33. European Commission. (2006). *Eurobarometer: Europeans and their languages. Special Eurobarometer 243, Public Opinion Analysis sector of the European Commission, Brussels.*
34. Krackhardt, D. (1987). Cognitive social structures. *Social Networks, 9*, 109–134.
35. Okada, A. (2011). Centrality of asymmetric social network: Singular value decomposition, conjoint measurement, and asymmetric multidimensional scaling. In S. Ingrassia, R. Rocci & M. Vichi (Eds.), *New perspectives in statistical modelling and data analysis* (pp. 219–227). Springer.
36. Wasserman, S., & Faust, K. (1994). *Social Network Analysis: Methods and applications.* Cambridge University Press.
37. Verzani, J. (2011). *Getting started with RStudio: An integrated development environment for R.* O'Reilly Media.
38. IBM-SPSS. (2019). *IBM-SPSS Categories 26.* IBM Corp.

Chapter 3
Analysis of Symmetry and Skew-Symmetry

3.1 Preliminaries

In many applications, proximities can be viewed as composed of a similarity (symmetric) component and a dominance/preference (non-symmetric) component that graphical representations can help to analyse.

A small example helps to illustrate how to exploit such a decomposition in real applications. In a stimulus identification experiment, a group of subjects are asked to evaluate whether two stimuli, presented one after the other, are the same or not. Seven stimuli labelled A, B, C, D, E, F and G are under study in the experiment, and the percentage of times the stimulus *i* (*j*) is "confused" with stimulus *j* (*i*) is reported for each pair *i,j* (*j,i*) in Table 3.1. For instance, when stimulus E (column) follows stimulus B (row), the confusion percentage is 22.8%, whereas when stimulus B (column) follows stimulus E (row), the confusion percentage is equal to 24.7%. The percentages of confusion of each stimulus with itself are also assessed and reported along the main diagonal of the table.

The data represent an example of a confusion matrix, which is a kind of asymmetric proximity where percentages measure similarities between stimuli.

In order to analyse the global level of confusion between each pair of stimuli, the averages of the percentages in cells (*i,j*) and (*j,i*) are computed and reported in Table 3.2.

Different levels of global confusion characterize the seven stimuli. For instance, stimulus B is much more confused with stimulus E (23.75%) than with stimulus C (11.4%), and stimulus D is much more confused with stimulus C (26.1%) than with stimulus F (14.25%). Percentages in the diagonal of the table show that stimuli are identified in a different manner, stimulus G is almost always identified (97%) while stimulus A is identified only 84% of the times.

A relevant property for the entries of Table 3.2 is that the percentage in cell (*i,j*) coincides with the percentage in cell (*j,i*), so that, if we denote by $m_{ij} = 1/2(\omega_{ij} + \omega_{ji})$ the average of the corresponding percentages ω_{ij} and ω_{ji} of the

G. Bove et al., *Methods for the Analysis of Asymmetric Proximity Data*,
Behaviormetrics: Quantitative Approaches to Human Behavior 7,
https://doi.org/10.1007/978-981-16-3172-6_3

Table 3.1 Percentages of confusion between seven stimuli

Stimuli	A	B	C	D	E	F	G
A	84.0	15.1	21.0	20.5	5.8	11.8	0.6
B	23.4	90.0	11.3	20.4	22.8	14.4	15.0
C	25.0	11.5	87.0	24.7	18.5	7.0	9.4
D	22.0	21.6	27.5	91.0	15.0	14.8	8.4
E	16.0	24.7	20.0	19.9	92.0	17.2	18.8
F	25.4	16.5	18.9	13.7	17.6	90.0	15.0
G	20.7	24.6	21.0	28.4	21.5	17.0	97.0

Table 3.2 Average percentages of the confusions in cells (i,j) and (j,i)

Stimuli	A	B	C	D	E	F	G
A	84.00	19.25	23.00	21.25	10.90	18.60	10.65
B	19.25	90.00	11.40	21.00	23.75	15.45	19.80
C	23.00	11.40	87.00	26.10	19.25	12.95	15.20
D	21.25	21.00	26.10	91.00	17.45	14.25	18.40
E	10.90	23.75	19.25	17.45	92.00	17.40	20.15
F	18.60	15.45	12.95	14.25	17.40	90.00	16.00
G	10.65	19.80	15.20	18.40	20.15	16.00	97.00

confusion matrix, then matrix $\mathbf{M} = (m_{ij})$ results symmetric $(\mathbf{M} = \mathbf{M}')$, that is, $m_{ij} = m_{ji}$ $(i,j = 1,2,\ldots,n)$.

Departures of the percentages ω_{ij} from the corresponding averages m_{ij}, reported in Table 3.3, can be analysed to detect the presence of systematic structures (non-random deviations from the averages). It is possible to notice that for some pairs of stimuli (e.g. stimuli B and C) deviations are small, which means that confusions are almost the same in both directions (e.g. B followed by C with confusion 11.3% and C followed by B with confusion 11.5%). Conversely, for other pairs of stimuli (e.g. stimuli A and G) deviations are large, which means that confusions are very different

Table 3.3 Departures from the averages of percentages of confusion

Stimuli	A	B	C	D	E	F	G
A	0.00	−4.15	−2.00	−0.75	−5.10	−6.80	−10.05
B	4.15	0.00	−0.10	−0.60	−0.95	−1.05	−4.80
C	2.00	0.10	0.00	−1.40	−0.75	−5.95	−5.80
D	0.75	0.60	1.40	0.00	−2.45	0.55	−10.00
E	5.10	0.95	0.75	2.45	0.00	−0.20	−1.35
F	6.80	1.05	5.95	−0.55	0.20	0.00	−1.00
G	10.05	4.80	5.80	10.00	1.35	1.00	0.00

in the two directions (e.g. A followed by G with confusion 0.6% and G followed by A with confusion 20.7%). Deviations along the main diagonal are null because they concern the same stimulus.

An important property for deviations in Table 3.3 is that entries in cells (*i,j*) and (*j,i*) have the same size but opposite sign. Thus, by denoting deviations by n_{ij}, it is not difficult to see that $n_{ij} = \omega_{ij} - m_{ij} = 1/2(\omega_{ij} - \omega_{ji})$, and matrix $\mathbf{N} = (n_{ij})$ results skew-symmetric ($\mathbf{N} = -\mathbf{N}'$), that is, $n_{ij} = -n_{ji}$ $(i,j = 1,2,\ldots,n)$. Moreover, the definition of deviation n_{ij} implies $\omega_{ij} = m_{ij} + n_{ij}$, and in matrix notation

$$\boldsymbol{\Omega} = \mathbf{M} + \mathbf{N}. \tag{3.1}$$

The additive decomposition (3.1) is general and means that any proximity (and more generally square) matrix can be decomposed into the sum of a symmetric matrix **M** and a skew-symmetric matrix **N**, with $m_{ij} = 1/2(\omega_{ij} + \omega_{ji})$ and $n_{ij} = 1/2(\omega_{ij} - \omega_{ji})$. It is not difficult to prove that decomposition (3.1) is unique (e.g. Borg & Groenen [1, pp. 496–497]), and that matrix **M** is the best symmetric approximation of matrix $\boldsymbol{\Omega}$ in the least-squares sense. For examples of other additive and multiplicative decompositions of proximities in the frequency domain, the reader is referred to Heiser and Busing [2].

The contribution of matrices **M** and **N** to the decomposition of the proximity matrix $\boldsymbol{\Omega}$ is not equal, as it is clear when entries in Tables 3.2 and 3.3 are compared. Deviations n_{ij} are usually much smaller than averages m_{ij} $(n_{ij} \leq m_{ij})$. The two contributions can be measured by computing the sum of squares of the elements of each component matrix compared to the sum of squares of the elements of the observed proximity matrix

$$\mathrm{SS}(\boldsymbol{\Omega}) = \sum_{i=1}^{n}\sum_{j=1}^{n}\omega_{ij}^2$$

$$\mathrm{SS}(\mathbf{M}) = \sum_{i=1}^{n}\sum_{j=1}^{n}m_{ij}^2$$

$$\mathrm{SS}(\mathbf{N}) = \sum_{i=1}^{n}\sum_{j=1}^{n}n_{ij}^2.$$

Constantine and Gower [3] remark that the following orthogonal breakdown of the sum of squares holds, reflecting the additive decomposition (3.1):

$$\mathrm{SS}(\boldsymbol{\Omega}) = \mathrm{SS}(\mathbf{M}) + \mathrm{SS}(\mathbf{N}). \tag{3.2}$$

So, the contributions of the symmetric and skew-symmetric components to the decomposition of the proximity matrix are given by the ratios

$$\frac{SS(\mathbf{M})}{SS(\boldsymbol{\Omega})}, \quad \frac{SS(\mathbf{N})}{SS(\boldsymbol{\Omega})}.$$

The two ratios are always non-negative and smaller than or equal to one, and the first ratio is always larger than or equal to the second ratio provided that proximities ω_{ij} are non-negative. The two ratios can be expressed as a percentage by multiplying them by one hundred. For the confusion matrix in Table 3.1, the contributions of the symmetric matrix **M** in Table 3.2 and the skew-symmetric matrix **N** in Table 3.3 are 96.1% and 3.9%, respectively. As usual, the symmetric component of the proximity is dominant with respect to the skew-symmetric component, and this is the reason why the asymmetry (supposed to be due to random error) has been disregarded in many applications and only the symmetric component of the proximities has been analysed. In this perspective, it has been also proposed to test the hypothesis of symmetry, for instance, in Bowker [4], Hubert and Baker [5] and Saburi and Chino [6]. However, as many studies have shown, the skew-symmetric component can reveal interesting information in spite of its small contribution to the additive decomposition (3.1).

The analysis of the two components is made easier by graphical representation, which can concern one or both components. The two components can be modelled and represented either separately or jointly. Separate representations are usually more informative, in particular for the skew-symmetric component, but they may be difficult to synthesize. Joint representations are easy to manage (e.g. spatial representations containing only *n* points) and equally fine to reveal some peculiar type of asymmetry. Zielman and Heiser [7] showed that many asymmetric models proposed in the literature can be analogously decomposed into a symmetric and a skew-symmetric part by opportune reparameterizations, making easier model interpretation in terms of similarity and preference.

In some applications, the symmetric component of the proximities is not interesting or it is not defined. Table 3.4 represents an example related to the 1997 score sheet of one of Japan's professional baseball leagues, the Central League [8]. In the table, any entry in the *(i,j)th* cell represents the number of games where Team *i* beats Team *j*. There were no games ended in a draw and for each pair of Teams *(i,j)*, $m_{ij} = 1/2(\omega_{ij} + \omega_{ji}) = 13.5$ holds.

Table 3.4 Score sheet of the Central League in 1997 [8]

Teams	1	2	3	4	5	6
1. Yakult	–	13	15	19	20	16
2. Yokohama	14	–	16	13	10	19
3. Hiroshima	12	11	–	13	12	18
4. Yomiuri	8	14	14	–	14	13
5. Hanshin	7	17	15	13	–	10
6. Chunichi	11	8	9	14	17	–

Thus, each entry of the symmetric component matrix **M** is 13.5, the diagonal entries are not defined, and all the relevant information regards the dominance relation retained by the skew-symmetric matrix **N**, that is, the only component worth to be analysed.

Another application domain can be the game balance data concerning financial players, where if player i beats player j, then i wins all the money that j loses. For such data, ω_{ij} are the gains of player i on player j, so that proximity matrix $\mathbf{\Omega}$ is skew-symmetric ($\omega_{ij} = -\omega_{ji}$) and the symmetric component is not defined.

Data regarding preference choices are another example of domain where the symmetric component is not very interesting. Preferences are *intrinsically asymmetric* because object (or stimulus) j cannot be preferred to object i if object i is preferred to object j. Frequently, preliminary transformations of ω_{ij} (e.g. number of times object i is preferred to object j) make preference data skew-symmetric, as for instance

$$\omega_{ij}^* = \log \frac{\omega_{ij}}{\omega_{ji}}$$

with matrix $\mathbf{\Omega}^* = \left(\omega_{ij}^*\right)$ skew-symmetric.

In the next section, some of the methods proposed to represent, separately or jointly, symmetry and skew-symmetry will be presented, for the analysis of confusion data and, more generally, of proximity data where the asymmetry is worth to be considered.

3.2 Separate Representation

The decomposition of the sum of squares in (3.2) proves that the symmetric and the skew-symmetric components are uncorrelated so that they can be viewed independently, and studied by separate models. In fact, a consequence of (3.2) is also that the least-squares approximation problem of matrix $\mathbf{\Omega}$ by a matrix

$$\mathbf{A} = \mathbf{S} + \mathbf{K},$$

where matrix **S** is symmetric and matrix **K** is skew-symmetric, may be split into two independent least-squares problems for the symmetric and skew-symmetric components **M** and **N**, respectively

$$\begin{aligned} LS(\mathbf{\Omega}, \mathbf{A}) &= \|\mathbf{\Omega} - \mathbf{A}\|^2 = \|\mathbf{M} - \mathbf{S}\|^2 + \|\mathbf{N} - \mathbf{K}\|^2 \\ &= LS_1(\mathbf{M}, \mathbf{S}) + LS_2(\mathbf{N}, \mathbf{K}), \end{aligned} \tag{3.3}$$

where the symbol $\|.\|$ represents the Frobenius norm of a matrix (defined in Chap. 2) or any other weighted least-squares approximation function (see Critchley [9] for a mathematical treatment). Hence, any model for matrix $\mathbf{\Omega}$ that can be written as the sum of a symmetric component ($\mathbf{S}$) and a skew-symmetric component ($\mathbf{K}$), with parameters for symmetry independent of parameters for skew-symmetry, can be fitted by two separate approximation problems as in the last term in (3.3).

An advantage of the separate analysis is that one is not committed to a completely rigid model at the outset, having the possibility to deal with the representation problem by adopting different kinds of models (e.g. spatial vs non-spatial) and different kinds of transformations (e.g. metric vs non-metric) for the two components.

Constantine–Gower's approach

Following the approach suggested in Constantine and Gower [3], the symmetric component $\mathbf{M}$ may be analysed by the inner product model presented in Chap. 2 or by a distance model as in symmetric multidimensional scaling, whereas the skew-symmetric component $\mathbf{N}$ may be independently analysed by the singular value decomposition.

Representation of the symmetric component

The inner product model:

$$m_{ij} = b_{ij}\left(\boldsymbol{x}_i, \boldsymbol{x}_j\right) + \varepsilon_{ij} = \sum_{s=1}^{r} x_{is}x_{js} + \varepsilon_{ij}, \tag{3.4}$$

or the distance model,

$$m_{ij} = d_{ij}\left(\boldsymbol{x}_i, \boldsymbol{x}_j\right) + \varepsilon_{ij} = \sqrt{\sum_{s=1}^{r} \left(x_{is} - x_{js}\right)^2} + \varepsilon_{ij}, \quad (i, j = 1, \ldots, n) \tag{3.5}$$

can be assumed for the symmetric component $\mathbf{M}$, according to the interpretation of the entries m_{ij} as similarities or dissimilarities, respectively. In fact, as it is shown in Sect. 1.2, it is always possible to pass from a similarity to a dissimilarity measure (and vice versa). For instance, when proximities are confusion percentages they can be converted into proximities measuring dissimilarities before the estimation of model (3.5) by the simple transformation

$$\omega_{ij}^{\text{dis}} = 100 - \omega_{ij}^{\text{sim}}.$$

Alternatively, their transformation can be incorporated in the least-squares criterion chosen, as described in Sect. 2.3.1.

Finally, it is worth to remark that the estimation of model (3.4) or (3.5) provides a map where entries m_{ij} are approximated by inner products or distances, respectively, but other geometrical choices such as trees and graphs are also possible (the interested reader is referred to Zielman & Heiser [7]; Carroll & Arabie [10]).

Representation of the skew-symmetric cmponent

The skew-symmetric component **N** can be analysed by SVD (already presented in a general formulation in Chap. 2). The particular form of the SVD of a skew-symmetric matrix is given in the following equation

$$\mathbf{N} = \mathbf{P\Gamma Q}' = \mathbf{P\Gamma J P}', \tag{3.6}$$

where **P** and **Q** are the matrices of the left and right singular vectors, respectively, $\mathbf{\Gamma} = \mathrm{diag}(\gamma_1, \gamma_1, \gamma_2, \gamma_2, \ldots)$ is the matrix of the singular values (if n is odd, the last diagonal element is zero) and **J** is a block diagonal matrix with 2×2 matrices $\begin{bmatrix} 0 & 1 \\ -1 & 0 \end{bmatrix}$ along the diagonal, and if n is odd, the last diagonal element set to one to maintain orthogonality. For the reader with mathematical background, we recall that this particular form of SVD, with singular values γ_s coming in pairs, is a consequence of the property to occur in conjugate imaginary pairs of the eigenvalues of a skew-symmetric matrix. The presence of matrix **J** in (3.6) makes the pairs of the right singular vectors (columns of matrix **Q**) corresponding to the same singular value, be a permutation and reflection of the corresponding left singular vectors (columns of matrix **P**). This means, for instance, that the first pair of left singular vectors $(\mathbf{p}_1, \mathbf{p}_2)$ corresponds to the pair of the right singular vectors $(\mathbf{q}_1 = \mathbf{p}_2, \mathbf{q}_2 = -\mathbf{p}_1)$.

It follows that in scalar notation the SVD in (3.6) for $(i,j = 1,\ldots,n)$ can be written as

$$n_{ij} = \gamma_1\left(p_{i1}p_{j2} - p_{j1}p_{i2}\right) + \gamma_2\left(p_{i3}p_{j4} - p_{j3}p_{i4}\right) + \cdots etc.$$

Now, if we limit our investigation to the first two singular vectors, the truncated two-dimensional SVD in scalar notation takes the form

$$n_{ij} = \gamma_1\left(p_{i1}p_{j2} - p_{j1}p_{i2}\right) + \varepsilon_{ij}. \tag{3.7}$$

The interpretation of model (3.7) is not in terms of inner product but in terms of area [11]. It is known from the analytical geometry that when points $P_i = (p_{i1}, p_{i2})$ and $P_j = \left(p_{j1}, p_{j2}\right)$ are plotted in a diagram (see Fig. 3.1), the absolute value of $\left(p_{i1}p_{j2} - p_{j1}p_{i2}\right)$ is twice the area of the triangle with vertices at the origin O, P_i and P_j. If the left singular vectors $(\mathbf{p}_1, \mathbf{p}_2)$ (or vectors $\sqrt{\gamma_1}\mathbf{p}_1, \sqrt{\gamma_1}\mathbf{p}_2)$ provide the coordinates for points P_i $(i = 1,\ldots,n)$ in a diagram, model (3.7) indicates that the areas of the triangles that the pairs of points form with the origin are approximately proportional to the sizes of the skew-symmetries, whose signs are given by the plane

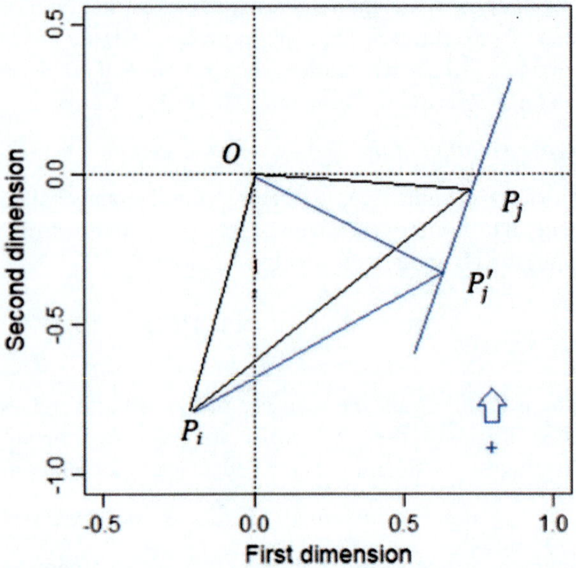

Fig. 3.1 Representation of skew-symmetry by singular value decomposition (*Gower diagram*)

orientation and depend on whether the vertices are traversed when you move from one (point) to the other in a clockwise or anticlockwise manner (positive anticlockwise in Fig. 3.1).

The goodness of the approximation depends on the proportion of SS(**N**) represented in the two dimensions of the diagram (sometimes called *Gower diagram* or *hedron* or *bimension*). As for bilinear models, the proportion of the variance of matrix **N** explained by model (3.7) can be computed by the ratio between the sum of squares of the singular values of the first *bimension* and the sum of squares of all the singular values.

Some properties of the diagram are worth to be mentioned because they can help graphical interpretation. Two close points P_i and P_j represent small values of skew-symmetry because of the small area of the corresponding triangle. Distant points approximately collinear with the origin represent also small skew-symmetry. Large skew-symmetries tend to be represented by pairs of points far from the origin and subtending at the origin angles of roughly 90 degrees. Points very close to the origin determine small triangle areas with any other point, and so they are usually characterized by small skew-symmetries.

All points P'_j on the line through P_j parallel to OP_i generate triangles $OP_iP'_j$ with the same areas as OP_iP_j (in Fig. 3.1, triangles OP_iP_j and $OP_iP'_j$ share the same base and height). So, each line parallel to OP_i is the locus of points equally skew-symmetric with P_i (this is the equivalent of a circle centred on P in a Euclidean space, that is, the locus of point equidistant from a point P). A minor problem is that (3.7) does not have a unique solution because any rigid rotation of the axes (vectors $\mathbf{p}_1, \mathbf{p}_2$) provides the same level of approximation. This problem does not influence the triangle areas however, so that diagram interpretation remains unchanged.

A more detailed description of the non-Euclidean geometry of such diagrams can be found in Gower [11, 12].

Application of the Constantine–Gower approach to Table 3.1

Now, we can show the results of the application of the methods for separate representations to the confusion matrix presented in Table 3.1. In this example, we used the program PROXSCAL available in IBM-SPSS, but we remark that many other MDS programs available in statistical packages could be used, as well as the routine available in the SMACOF R package (for further details, see the software section).

Percentages of confusion are assumed to be similarities, and for this reason the computational algorithm was allowed to perform also an optimal interval transformation in the process of least-squares minimization. The optimal interval transformation for this example results to be: $\hat{d}_{ij} = -0.067\ m_{ij} + 2.142$. As a result, two points representing stimuli are close when the level of the corresponding average confusion m_{ij} is large. The approximation of the distances to the transformed percentages (estimated *disparities* $\hat{d}_{ij}$) is acceptable, the value of *Stress*-I is 0.18. In Fig. 3.2a, the seven stimuli of Table 3.1 marked in letters are represented in two dimensions.

The positions of the points are informative on the averages of the percentages of confusion in Table 3.2. In fact, stimulus B is near stimulus E because they have large confusion (23.75%), and the same happens for stimuli C and D (26.1%). Conversely, stimulus A is located far from stimulus G because their percentage of confusion is low (10.65%), as well as stimulus B is far from stimulus C (11.4%). However, not all the percentages are well represented in the diagram. For instance, according to the entries of Table 3.2, stimulus F should be a bit closer to stimulus G (16%) than to stimulus B (15.45%).

Figure 3.2b presents the diagram obtained by applying model (3.7) to the skew-symmetric matrix in Table 3.3, whose entries are the departures from the averages of the percentages of confusion (already represented in Fig. 3.2a). The model can be estimated by a routine available in the Asymmetry R package (see the software section for details). The proportion of variance accounted for by model (3.7) is 88.3% of the total variance of the skew-symmetric matrix, and the coordinates of the points in the diagram are provided by the two singular vectors ($\mathbf{p}_1, \mathbf{p}_2$) in Table 3.5.

Since the corresponding singular value γ_1 is 18.82, for instance, the approximation of model (3.7) for the skew-symmetry between stimuli A and B (-4,15%) is [$18.82 \times (0.72 \times (-0.2) - 0.25 \times 0.0) = -2.71$], and for the skew-symmetry between stimuli A and G ($-10{,}05$%) is [$18.82 \times (0.72 \times (-0.81) - (-0.23 \times 0.0) = -10.98$]. In the diagram, the areas of the triangles that pairs of points representing stimuli form with the origin, multiplied by $2\gamma_1$ ($2 \times 18.82 = 37.6$), approximate the corresponding skew-symmetries (or multiplied by 2 if the coordinate vectors are $\sqrt{\gamma_1}\mathbf{p}_1, \sqrt{\gamma_1}\mathbf{p}_2$).

Stimuli A and G form the triangle with the largest area, because the corresponding skew-symmetry in Table 3.3 has the highest absolute value. The sign of the skew-symmetry is positive counterclockwise (from G to A), so the entry in row G and column A is positive and the entry in row A and column G is negative. This means also that confusion is larger when stimulus A is presented after stimulus G than the

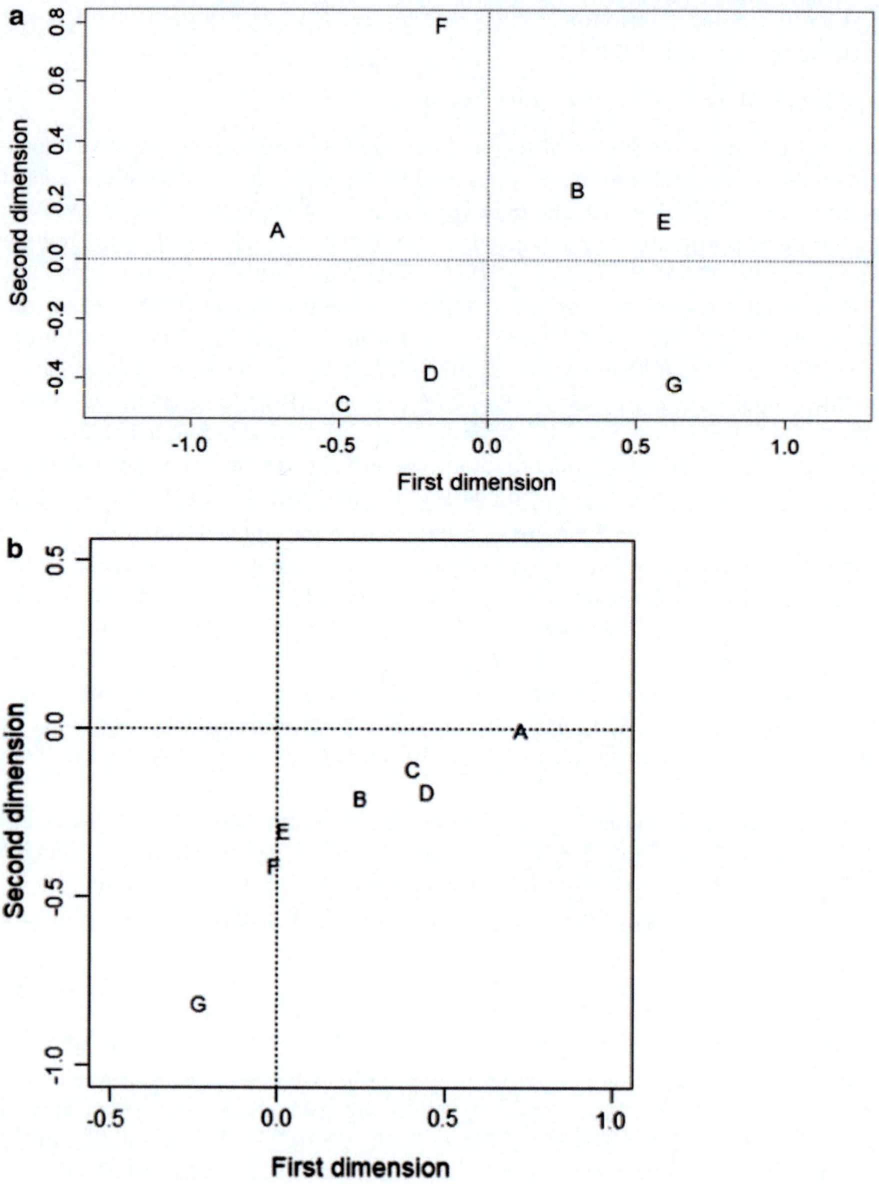

Fig. 3.2 **a** Distance representation of data in Table 3.2 and **b** representation of skew-symmetry of Table 3.3 by singular value decomposition

Table 3.5 Coordinates of points representing stimuli in Fig. 3.2b

Stimuli	$\mathbf{p}_1$	$\mathbf{p}_2$
A	0.72	0.00
B	0.25	−0.20
C	0.40	−0.12
D	0.44	−0.18
E	0.02	−0.30
F	-0.002	−0.40
G	-0.23	−0.81

reverse. Thus, stimulus G is judged more similar to stimulus A than the reverse, which means that stimulus A is a *prototype* (i.e. the more salient or the referent) stimulus with respect to stimulus G, considered a *variant* (i.e. less salient) of A (e.g. we say "the son resembles the father" rather than "the father resembles the son"). Points representing stimuli C and D are very close, so the area of the triangle they form with the origin is small as their skew-symmetry, that is, positive (1.4%) when stimulus C follows stimulus D. Similar comments can be done for stimuli E and F that present very small skew-symmetry.

3.2.1 *One-dimensional Skew-Symmetry*

An interesting aspect of the skew-symmetries of Table 3.3 is that the points representing stimuli are roughly aligned on a line (with the exception of point G) in the diagram in Fig. 3.2b. Hence, the skew-symmetry could also be approximated on one dimension to make interpretation easier than in Fig. 3.2b, and model (3.7) could be simplified in the model

$$n_{ij} = \gamma_1 \left(p_{i1} - p_{j1} \right) + \varepsilon_{ij}, \tag{3.8}$$

that can be obtained as a special case when $\mathbf{p}_2$ is set equal to a vector of ones in (3.7) ($p_{i2} = p_{j2} = 1$). Hence, setting $r_h = \gamma_1 p_{h1}$ (h = 1,2,...,n), model (3.8) can be written

$$n_{ij} = \left(r_i - r_j \right) + \varepsilon_{ij}. \tag{3.9}$$

A unique least-square solution for model (3.9) cannot be identified, because, given the solutions $\hat{r}_i$, any translation $\hat{r}_i + c$ by any constant c represents equivalent solution. Without loss of generality, if the constraint $\sum_{i=1}^{n} r_i = 0$ is assumed, it is easy to prove that least-squares estimates of r_i (i = 1,...,n) are the averages of the n_{ij} values within each row i over the columns, that is,

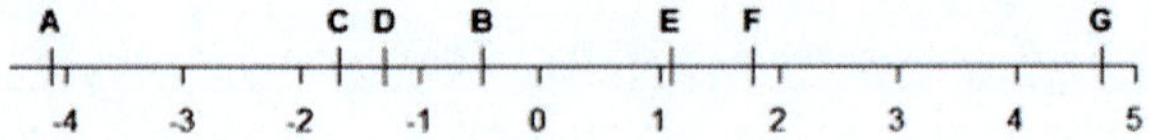

Fig. 3.3 Representation of skew-symmetry of Table 3.3 on one dimension by model (3.9)

$$\hat{r}_i = \frac{1}{n}\sum_{j=1}^{n} n_{ij}. \tag{3.10}$$

In the example of the seven stimuli, we can easily obtain $(\hat{r}_1, \hat{r}_2, \ \ldots, \hat{r}_7) = (-4.12, -0.48, -1.68, -1.31, 1.1, 1.78, 4.71)$ by Table 3.3. Hence, stimuli can be represented along one dimension as in Fig. 3.3 and the skew-symmetries can be analysed as differences between the estimates $\widehat{r_i}$ (oriented or signed Euclidean distances) instead of triangle areas as in Fig. 3.2b.

The proportion of variance accounted for by model (3.9) is 84.4%, only four percentage points less than the corresponding proportion for model (3.7), so that the levels of approximation of the skew-symmetry for the two models are not very different. For some cells of the table, the simple unidimensional model (3.9) performs better than model (3.7). For instance, the skew-symmetry between stimuli A and B (-4.15%) is approximated by model (3.9) by computing the oriented distance between A and B, that is, the difference $[(\hat{r}_1 - \hat{r}_2) = -4.12\text{-}(-0.48) = -3.64\%]$. In other cases, the general model performs better than the simple model, for instance: the skew-symmetry between stimuli A and G (-10,05%) is approximated by model (3.9) by computing the difference $[(\hat{r}_1 - \hat{r}_7) = \text{-}4.12\text{–}4.71 = -8.84\%]$.

A disadvantage of model (3.9) is that does not allow to represent circular triads (for the definition of circular triad, see Sect. 1.2—*Structures for proximity data*).

3.2.1.1 Models Linked to One-dimensional Skew-Symmetry*

In this section, some models are briefly described because they present interesting relationships with one-dimensional skew-symmetry. References are provided for further study.

Weeks and Bentler's model

One-dimensional skew-symmetry has been also incorporated in the following model proposed by Weeks and Bentler [13]

$$\omega_{ij} = \alpha \ d_{ij}\left(\boldsymbol{x}_i, \boldsymbol{x}_j\right) + \beta + \left(r_i - r_j\right) + \varepsilon_{ij}, \tag{3.11}$$

where $d_{ij}\left(\boldsymbol{x}_i, \boldsymbol{x}_j\right)$ is the Euclidean distance, α and β are two parameters that allow for a linear relation between proximities and distances (in particular α can be set to -1 when proximities measure similarities), r_i and r_j represent the skew-symmetric component and ε_{ij} is the error term. An iterative Gauss–Newton algorithm to obtain least-squares estimates of the parameters was proposed in Weeks and Bentler [13].

Similar to the Constantine and Gower's approach, the estimated distances allow to represent the symmetric component in a (usually two-dimensional) map and the estimates of parameters r_i, represented in a separate diagram as in Fig. 3.3, provide information on the skew-symmetric component. Later in this chapter, it will be shown that a special case of model (3.11) termed radius–distance model allows for a joint representation of the two components **M** and **N**. Extensions of the model to deal with several proximity matrices (three-way case) are considered in Chap. 5.

Quasi-independence and quasi-symmetry models

The quasi-independence and quasi-symmetry models for square tables presented in Caussinus and de Falguerolles [14] are closely connected to Weeks and Bentler's model. For proximities measuring similarity relations at an interval level scale (e.g. ratings), two additive models are considered ($i \neq j$): the *quasi-independence model*

$$\omega_{ij}^{\text{sim}} = \beta + r_i + c_j + \varepsilon_{ij}^*, \quad (i \neq j)$$

with constraints $\sum_i r_i = \sum_j c_j = 0$, and the *quasi-symmetry model*

$$\omega_{ij}^{\text{sim}} = s_{ij} + \beta + r_i + c_j + \overset{\circ}{\varepsilon}_{ij} \quad \big(s_{ij} = s_{ji}\big)(i \neq j),$$

with constraints $\sum_{i,i\neq j} s_{ij} = \sum_{j,j\neq i} s_{ij} = \sum_i r_i = \sum_j c_j = 0$; residuals ε_{ij}^* and $\overset{\circ}{\varepsilon}_{ij}$ are assumed independent and identically distributed (or *i.i.d.*, a standard hypothesis in statistical models). Parameter β represents a general mean proximity value, r_i a row effect and c_j a column effect; parameters s_{ij} reflect the similarities between rows and columns. Least-squares estimates of parameters are provided in Caussinus and de Falguerolles [14, p. 39], with the following interesting properties:

$$\hat{s}_{ij} = \frac{\widehat{\varepsilon}_{ij}^* + \widehat{\varepsilon}_{ji}^*}{2}$$

$$\widehat{\overset{\circ}{\varepsilon}}_{ij} = \frac{\widehat{\varepsilon}_{ij}^* - \widehat{\varepsilon}_{ji}^*}{2},$$

and so

$$\widehat{\varepsilon}_{ij}^* = \hat{s}_{ij} + \widehat{\overset{\circ}{\varepsilon}}_{ij}.$$

Graphical representation can help the analysis and interpretation of the previous models in many ways. For instance, residuals $\widehat{\varepsilon}_{ij}^*$ can be represented in a biplot and residuals $\widehat{\overset{\circ}{\varepsilon}}_{ij}$ in a Gower diagram, as suggested in Caussinus and de Falguerolles [14]. Row and column effects can be compared in a Cartesian diagram, and appropriate transformations of the symmetric parameters $\hat{s}_{ij}$ can be represented by symmetric multidimensional scaling [14, pp. 42–43].

Weeks and Bentler's model can be considered as a special case of the quasi-symmetry model when $s_{ij} = \alpha d_{ij}(\boldsymbol{x}_i, \boldsymbol{x}_j)$ and $c_j = -r_j$. Moreover, we remark that graphical representations of the parameters of the quasi-symmetry models have been also considered for square contingency tables (e.g. mobility tables) by Caussinus and de Falguerolles [14, pp. 45–50] and many others (for a review, the interested reader can refer to Zielman & Heiser [7]; de Rooij & Heiser [15, 16]; de Rooij [17]).

Similarity and bias model

In addition, the previous quasi-symmetry model is a special case of a general class of models termed *similarity and bias model* [18, 19]. Following the model, the proximity ω_{ij} can be expressed as

$$\omega_{ij} = F\left(s_{ij}^* + r_i^* + c_j^*\right) + \varepsilon_{ij}, \tag{3.12}$$

where F is a general monotonic function, s_{ij}^* is a symmetric similarity function, r_i^*, c_j^* are row and column bias functions and ε_{ij} is the error term. Choosing F as the identity transformation and setting $s_{ij}^* = s_{ij} + \beta$, $r_i^* = r_i$ and $c_j^* = c_j$, the quasi-symmetry model is obtained as a special case of (3.12). Other models with different types of structures (e.g. additive tree) assumed for the similarity function have been proposed, and they are described in detail in Zielman and Heiser [7]. A non-distance model which is a special case of (3.12) is also proposed in Brossier [20] for clustering a set of objects (see Chap. 4).

3.2.2 Distance Models for Skew-Symmetry*

Since judgements based on distances are cognitively less demanding than judgements based on triangle areas, methods based on distance models were also considered to make easy the diagram interpretation of the skew-symmetry.

First of all, we note that any skew-symmetric matrix contains two types of information for each pair (*i,j*): size (or magnitude) and sign (or directionality), given by the absolute value $|n_{ij}|$ and $sign(n_{ij})$, respectively. So, each entry n_{ij} of matrix $\mathbf{N}$ can be written as

$$n_{ij} = sign(n_{ij})|n_{ij}|, \tag{3.13}$$

where $|n_{ij}|$ is the absolute value of n_{ij}, $sign(n_{ij}) = 1$ if $n_{ij} > 0$, $sign(n_{ij}) = -1$ if $n_{ij} < 0$, $sign(n_{ij}) = 0$ if $n_{ij} = 0$.

A simple method for graphical representation of matrix $\mathbf{N}$ was proposed by Bove [21], adapting an idea proposed by Okada and Imaizumi [22] for asymmetric proximities (presented later in this chapter) to skew-symmetric data. The method is based on the following two-step procedure.

STEP 1.

The sizes of skew-symmetries $|n_{ij}|$ are symmetric $(|n_{ij}| = |n_{ji}|)$, so matrix $\mathbf{T} = (t_{ij}) = (|n_{ij}|)$ can be considered as a dissimilarity matrix that can be approximated by distances in a *low-dimensional* Euclidean space (usually two-dimensional) according to the following model:

$$f(t_{ij}) = d_{ij}(\boldsymbol{x}_i, \boldsymbol{x}_j) + \varepsilon_{ij} = \sqrt{\sum_{s=1}^{r} (x_{is} - x_{js})^2} + \varepsilon_{ij} \tag{3.14}$$

where f is the transformation function, $d_{ij}(\boldsymbol{x}_i, \boldsymbol{x}_j)$ is the distance between objects i and j, x_{is} and x_{js} are the coordinates on dimension s of i and j, respectively, and ε_{ij} is the error component. The model can be easily estimated by standard statistical software containing symmetric multidimensional scaling routines. An advantage of this model is that it is easy to incorporate both non-metric approaches and external information regarding the objects.

STEP 2.

The signs of skew-symmetries $sign(n_{ij})$, collected in matrix $\mathbf{K}^{\text{sign}} = \left(k_{ij}^{\text{sign}}\right) = (\text{sign}(n_{ij}))$, can be approximated by the model

$$k_{ij}^{\text{sign}} = (r_i - r_j) + \varepsilon_{ij} \ , \tag{3.15}$$

already considered for matrix $\mathbf{N}$ in Eq. (3.9). According to (3.10), least-squares estimates of the parameters r_i in this case are

$$\hat{r}_i = \frac{1}{n} \sum_{j=1}^{n} k_{ij}^{\text{sign}}. \tag{3.16}$$

Graphical representation

Coordinates x_{is} provided by model (3.14) allow us to represent in a plane ($r = 2$) the sizes of the skew-symmetries t_{ij} by the distances d_{ij} (the larger the distance the larger the size). The estimates $\hat{r}_i$ in (3.16) can be used to draw circles around the points with coordinates (x_{i1}, x_{i2}) obtained in Step 1. The radii of the circles can be obtained from the non-negative estimates $\hat{r}_i$ by an appropriate translation by a constant c such that $min(\hat{r}_i + c) = 0$ (the smallest radius is equal to zero, and radii can be also rescaled to be easily comparable). In order to interpret the graphical representation in the plane, we remark that two points i,j far from (close to) each other represent a large (small) skew-symmetry, and when the circle around point i is larger than circle around point j, then the estimate of the skew-symmetry n_{ij} is positive and the estimate of the skew-symmetry n_{ji} is negative. An R algorithm for the two-step method is provided in the software section.

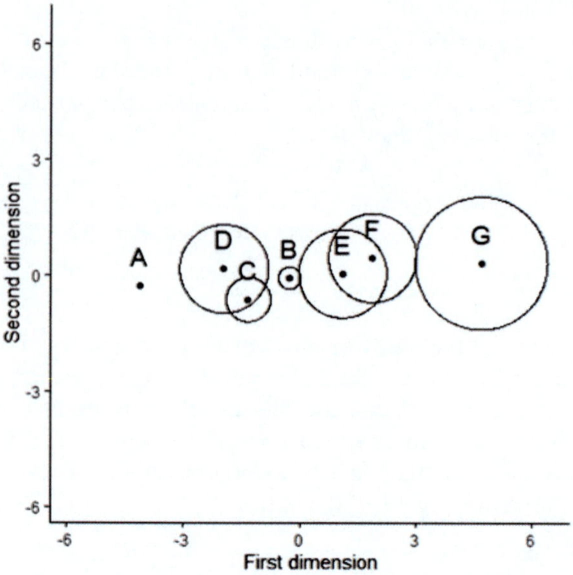

Fig. 3.4 Representation of skew-symmetry of Table 3.3. The distance between points represents the size of the skew-symmetry, and the difference between circles represents the sign of the skew-symmetry

The application of the method to the skew-symmetric matrix in Table 3.3 provides the graphical representation of Fig. 3.4. The alignment of the points representing stimuli already detected in Fig. 3.2b is quite evident. Stimuli A and G show the largest distance because their skew-symmetry has the largest size, as already noted in Fig. 3.2b. The circle centred in G is larger than circle centred in point A; this means that skew-symmetry from A to G is negative, whereas skew-symmetry from G to A is positive. Stimuli C and D are very close, which means that the size of their skew-symmetry is small, and comparing the corresponding circles we can say that is positive when stimulus C follows stimulus D. Therefore, the smaller the size of the circle, the more the stimulus can be considered as a prototype. In summary, the analysis of Fig. 3.4 provides the same information obtained from Fig. 3.2b but in a more direct way, because distances and circles are compared instead of triangle areas and positions of the points in (counter)clockwise direction. A disadvantage of the two-step procedure with respect to model (3.7) is that two separate levels of approximations have to be considered, one for model (3.14) and the other for model (3.15). For the first approximation, the value of Stress-I is 0.35, while for the second one the percentage of signs correctly represented should be preferred with respect to least-squares approximation of model (3.15) (more than 70% of the signs of the skew-symmetries have been correctly represented). Another possible disadvantage is that the two-step procedure is not able to represent circular triads.

On the same line, Borg and Groenen [1] proposed the following distance model for skew-symmetry

$$n_{ij} = sign\big(\boldsymbol{x}_i'\mathbf{T}\boldsymbol{x}_j\big)\, d_{ij}\big(\boldsymbol{x}_i, \boldsymbol{x}_j\big) + \varepsilon_{ij},$$

where $\boldsymbol{x}_i, \boldsymbol{x}_j$ are the vectors of coordinates for objects i and j in $r = 2$ dimensions, $T = \begin{pmatrix} 0 & -1 \\ 1 & 0 \end{pmatrix}$, $d_{ij}(\boldsymbol{x}_i, \boldsymbol{x}_j)$ is the Euclidean distance between objects i and j and ε_{ij} is the error component. The distances between points estimate the sizes of the skew-symmetries, and the directions of the rotations provide their signs. If the angle measured clockwise between the vector to point i and the vector to point j is less than 180°, then n_{ij} is estimated by $d_{ij}(\boldsymbol{x}_i, \boldsymbol{x}_j)$. Conversely, if this angle is between 180° and 360° measured clockwise (or, equivalently, between 0 and − 180° in the counterclockwise sense), then n_{ij} is estimated by $-d_{ij}(\boldsymbol{x}_i, \boldsymbol{x}_j)$. Borg and Groenen [1, pp. 501–502] also fit the model to the largely known Morse-code confusion data to show its performance. For the least-squares estimation, Borg and Groenen [1] applied a general-purpose optimization routine that can be quite sensitive to local optima.

3.3 Joint Representation

The graphical display of symmetry and skew-symmetry can be combined in informative ways to grasp more easily how the two components are related to each other. The joint representation can be obtained with models and methods with parameters for symmetry and skew-symmetry either independent or dependent.

3.3.1 Independent Modelling of Symmetry and Skew-Symmetry

In the two-step procedure for distance representation of the skew-symmetry presented in Sect. 3.2.2, stimuli were displayed as circles with different radii in a plane, so that sizes and signs of the skew-symmetries were represented as distances and differences between radii, respectively. The idea to represent objects as points and circles (or spheres) with different radii in a plane (or in a space) is due to Okada and Imaizumi [22], who apply this approach to a proximity matrix, assuming an approximately one-dimensional skew-symmetry.

The model is presented here with a slightly different formalization with respect to Okada and Imaizumi [22] and Sect. 5.2.2.1, to make the comparison with the other models already considered in this section easy.

The model proposed is the *radius–distance model*

$$\omega_{ij}^{\text{dis}} = m_{ij} + n_{ij} = d_{ij}(\boldsymbol{x}_i, \boldsymbol{x}_j) + (r_i - r_j) + \varepsilon_{ij}, \tag{3.17}$$

where the symbols have been already defined. Any object is depicted as a point with coordinates $\boldsymbol{x}_i$, and it is the centre of a circle with radius r_i. The proximity ω_{ij}^{dis}

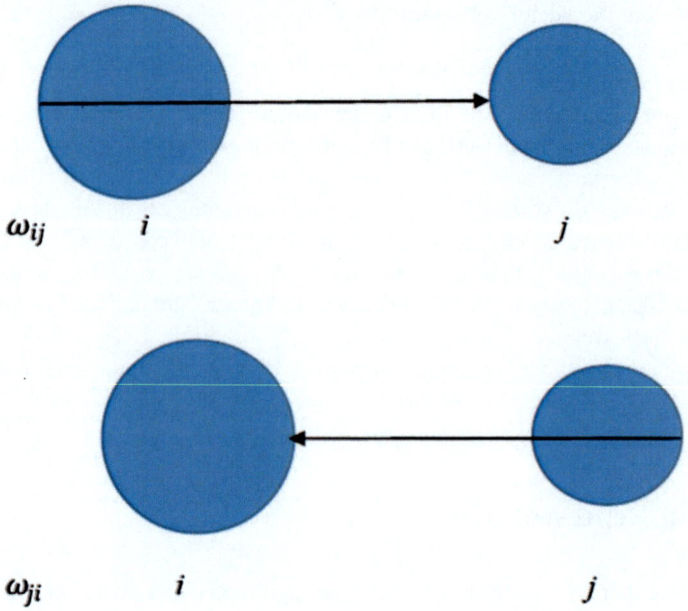

Fig. 3.5 Representation of the *radius–distances* between object i and object j

is modelled by the *radius–distance* from i to j, given by the distance $d_{ij}(\boldsymbol{x}_i, \boldsymbol{x}_j)$ between the centres of the circles, by adding the radius of object i and by subtracting the radius of object j (see Fig. 3.5 for this interpretation).

The least-squares approximation of matrix $\boldsymbol{\Omega}$ according to the decomposition (3.3) splits into the two following independent least-squares problems because parameters $\boldsymbol{x}_i$ are independent of parameters r_i,

$$LS(\mathbf{M}, \mathbf{D}) = \sum_{i=1}^{n} \sum_{j=1}^{n} (m_{ij} - d_{ij})^2$$

and

$$LS(\mathbf{N}, \mathbf{R}) = \sum_{i=1}^{n} \sum_{j=1}^{n} (n_{ij} - r_{ij})^2$$

with $\mathbf{D} = (d_{ij}) = (d_{ij}(\boldsymbol{x}_i, \boldsymbol{x}_j))$, $\mathbf{R} = (r_{ij}) = ((r_i - r_j))$. The former is a standard symmetric multidimensional scaling problem from which we obtain the coordinates of the objects, while the latter has the constrained analytical solutions $\hat{r}_i$ (3.10), from which the lengths of the radii are derived by the appropriate translation c attaining $min(\hat{r}_i + c) = 0$.

For the proximities assumed to measure similarities, the first least-squares problem is modified in order to incorporate an optimal linear transformation, as done in the representation of the symmetric component of Table 3.2. Thus, the average proximities m_{ij} are transformed by the optimal linear relation estimated by the minimization algorithm: $\hat{d}_{ij} = -0.067\, m_{ij} + 2.142$. Analogously, the lengths of the radii are rescaled by the same constant 0.067, in order to be compared with distances.

Figure 3.6 shows the diagram obtained by the estimation of the radius-distance model fitted to data of Table 3.1.

The symmetric component can be analysed by looking at the distances between the centres of the circles (large distances represent low levels of the mean percentages of confusion), so the same interpretation provided for the configuration in Fig. 3.2a holds. For instance, stimulus A is located far from stimulus G because their mean percentage of confusion is low (10.65%); conversely, stimulus B is close to stimulus E because they have a large mean percentage of confusion (23.75%).

Differences in circle sizes affect the sizes and the signs of the symmetry and not only the signs of the skew-symmetry as in Fig. 3.4. For instance, since the circle of stimulus G is larger than the circle of stimulus A (radius equal zero), the radius-distance from G to A is larger than the radius-distance from A to G (see Fig. 3.5 for this interpretation), which means that confusion is larger when stimulus A is presented after stimulus G than the reverse. Hence, stimuli with small circles could be interpreted as prototype stimuli in Fig. 3.6.

Okada and Imaizumi [22] and Borg and Groenen [1, p. 513] presented algorithms for a non-metric version of the radius–distance model. Okada [23] discussed an extension of the radius-distance model where each stimulus is represented as a point and an ellipse (ellipsoid, hyperellipsoid) in a multidimensional Euclidean space. The introduction of the ellipse to represent the stimulus has the advantage to allow for a better approximation of the skew-symmetry with respect to the distance radius

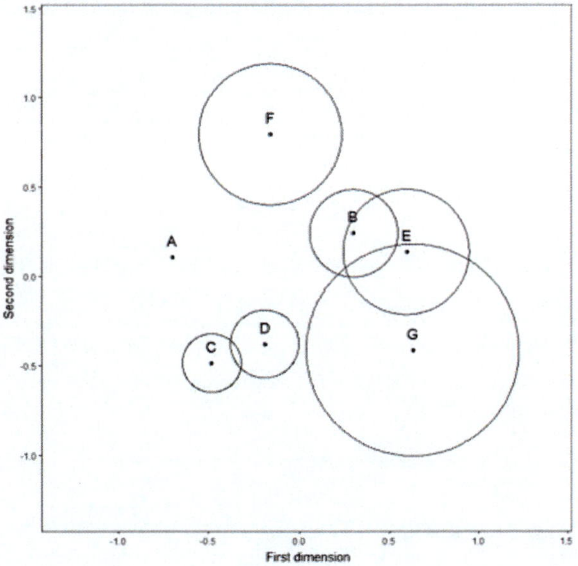

Fig. 3.6 Representation of the *radius–distances* for data in Table 3.1

model and to obtain a unique orientation of the axes of the stimulus configuration. Saburi and Chino [6] proposed a maximum likelihood approach for the estimation of the radius–distance model. Extensions to the three-way case will be presented in Chap. 5.

3.3.2 Dependent Modelling of Symmetry and Skew-Symmetry

Drift vector method

A method for displaying simultaneously the symmetric part and the skew-symmetric part of the proximity data is presented in Borg and Groenen [1]. Firstly, a map for the symmetric component matrix **M** is provided by a symmetric MDS method, as in the Constantine and Gower's approach. Then, the skew-symmetric information in **N** is incorporated in the map by drawing arrows from each point i to any other point j of the map, so that lengths and directions correspond to the entries in the rows of the skew-symmetric matrix **N**. To clarify how the procedure works, let us consider confusion data in Table 3.1. A map for the symmetric component **M** is already provided in Fig. 3.2a from symmetric MDS. Now, by considering, for instance, the row of stimulus G and the column of stimulus A in Table 3.3, an arrow will be attached at G pointing towards A. The arrow direction towards A means that stimulus G is more often confused with stimulus A when presented as first (positive skew-symmetry). The arrow direction goes from the variant stimulus to the prototype. Analogously, an arrow of the same length will be attached at point A, oriented in the opposite direction to point G (negative skew-symmetry). The two arrows are represented in Fig. 3.7,

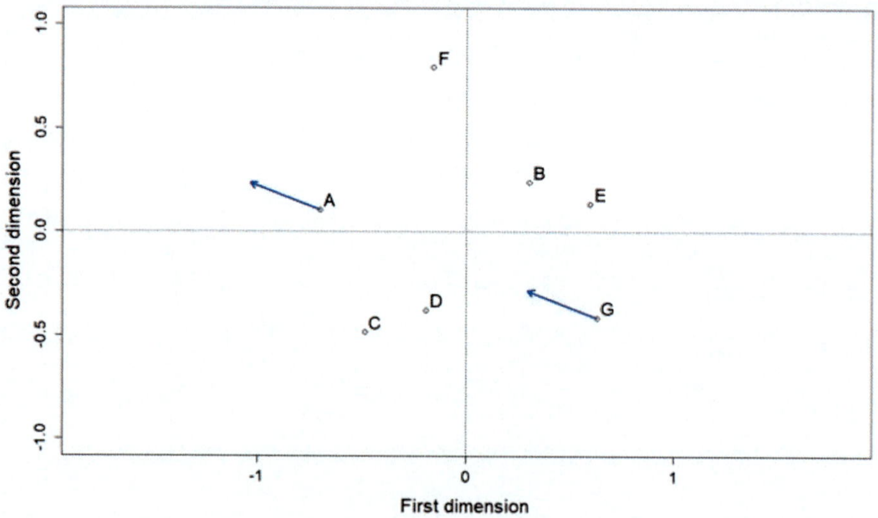

Fig. 3.7 Representation of arrows for the pair of stimuli A, G

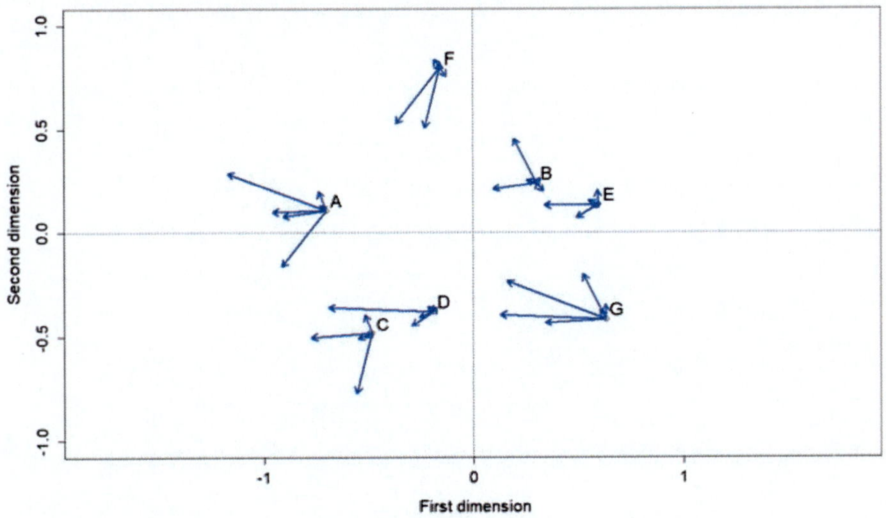

Fig. 3.8 Representation of arrows for the seven stimuli of Table 3.1

and the length of both arrows is chosen as $\gamma |n_{ij}|$, with γ a positive overall scaling factor (e.g. $1/mean(\omega_{ij})$) that allows to represent vectors most conveniently in the map.

All entries of matrix **N** can be represented as arrows in the same map by the procedure applied to stimuli A and G, and the resulting map is provided in Fig. 3.8.

The interpretation of the distances in the map is the same of Fig. 3.2a, so that stimulus A is located far from stimulus G because their mean percentage of confusion is low (10.65%); stimulus C is near stimulus D because of their large mean percentage of confusion (26.1%). Since the lengths of the arrows represent the sizes of the skew-symmetries, so the two arrows along the direction from G to A have large lengths because they represent a large skew-symmetry between stimuli A and G (10.05%). The two arrows along the direction from D to C are very short because they represent a small skew-symmetry (1.4%), and D dominates *C*.

When the number of rows (e.g. stimuli) of the proximity matrix is large, reporting all arrows can provide cluttered pictures, and it can be convenient to draw only the resultant (average) of the arrow bundle attached to each point. Even if the possibility to depict pairwise skew-symmetries is lost, the average arrow (called *drift vector*) can allow to better detect how the symmetric and the skew-symmetric data components are related to each other. In Fig. 3.9, the drift vectors for the seven stimuli are shown. The measurement unit has been chosen to better visualize small vectors. Most vectors point to the same direction from the right to the left side of the map, showing a systematic trend. Stimuli on the right side (B, E, G) are more confused when presented as first than the stimuli on the left side (A, C, D). The linearity of the skew-symmetry seems difficult to detect in Figs. 3.8 and 3.9, because it requires the pairwise comparisons between all the arrows in Fig. 3.8.

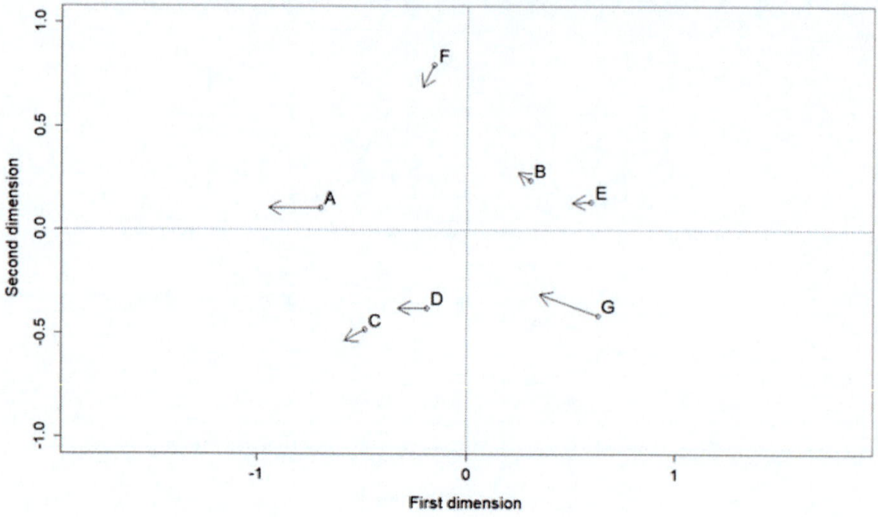

Fig. 3.9 Representation of the drift vectors for the seven stimuli of Table 3.1

The drift vectors can be obtained by the function *driftVectors* available in the SMACOF R package [24]. The arrows provided in Figs. 3.8 and 3.9 were computed by an "ad hoc" *R* program provided in the software section.

Chino–Escoufier–Grorud approach.

Following Constantine and Gower [3], a separate analysis of symmetric and skew-symmetric components is performed by considering distance (or inner product) models and SVD, respectively, with independent sets of parameters for the two components. Chino [25] and Escoufier and Grorud [26] independently took the different perspectives to explore the possibility to jointly represent the symmetric and skew-symmetric components with the same set of parameters, to possibly analyse both symmetry and skew-symmetry in the same configuration of points.

The model proposed by Escoufier and Grorud [26] in two dimensions can be formulated as

$$\omega_{ij} = m_{ij} + n_{ij} = (x_{i1}x_{j1} + x_{i2}x_{j2}) + \left(x_{i1}x_{j2} - x_{j1}x_{i2}\right) + \varepsilon_{ij}, \tag{3.18}$$

where $(x_{i1},\ x_{i2})$ are the coordinates of object i in a plane and ε_{ij} is the error component. In matrix notation, according to the decomposition (3.3), the least-squares criterion for model (3.18) can be written

$$\begin{aligned} LS(\mathbf{\Omega}\ , \mathbf{X}) &= LS_1(\mathbf{M}, \mathbf{X}) + LS_2(\mathbf{N}, \mathbf{X}) \\ &= \left\|\mathbf{M} - \mathbf{XX}'\right\|^2 + \left\|\mathbf{N} - \mathbf{XJX}'\right\|^2 \end{aligned}$$

$$= \sum_{i=1}^{n}\sum_{j=1}^{n}(m_{ij} - (x_{i1}x_{j1} + x_{i2}x_{j2}))^2$$

$$+ \sum_{i=1}^{n}\sum_{j=1}^{n}(n_{ij} - (x_{i1}x_{j2} - x_{j1}x_{i2}))^2, \tag{3.19}$$

where **X** is the $n \times 2$ matrix of coordinates and **J** the 2×2 matrix $\begin{bmatrix} 0 & 1 \\ -1 & 0 \end{bmatrix}$. It follows that in the configuration, the symmetric component m_{ij} is represented by the inner product of the vectors associated with points i and j, and the skew-symmetric component is represented by twice the area of the triangle having the two points i and j and the origin as vertices, whose algebraic sign is determined by the orientation of the plane (as in the Gower diagram). The lengths of the vectors represent the diagonal entries of the proximity matrix. The GIPSCAL model proposed by Chino [25, 27] has substantially the same graphical interpretation of model (3.18) with some additional parameters to differently weigh the two components and an additive constant to take into account the proximities defined at an interval level of measurement.

Escoufier and Grorud [26] showed that the least-squares estimate of the configuration matrix **X** is obtained by the first two eigenvalues and eigenvectors of the composite matrix

$$\mathbf{E} = \begin{pmatrix} \mathbf{M} & -\mathbf{N} \\ \mathbf{N} & \mathbf{M} \end{pmatrix}.$$

The estimation of model (3.18) can be achieved by any program containing routines for computing eigenvalues and eigenvectors (or SVD), and a simple *R* program is also provided in the software Sect. 3.6.

A measure of the approximation error (variance not accounted for or unexplained proportion of variance) of model (3.18) can be obtained (analogously to stress measures in multidimensional scaling) by the ratio of $LS\left(\mathbf{\Omega}, \widehat{\mathbf{X}}\right)$ to the norm of $\mathbf{\Omega}$, and for matrices **M** and **N** by the ratios of $LS_1\left(\mathbf{M}, \widehat{\mathbf{X}}\right)$ and $LS_2\left(\mathbf{N}, \widehat{\mathbf{X}}\right)$ to the norms of **M** and **N**, respectively (where $\widehat{\mathbf{X}}$ denotes the estimated configuration in two dimensions).

Figure 3.10 shows the configuration obtained from model (3.18) for the confusion matrix of Table 3.1. The quite large approximation errors for the symmetric and skew-symmetric components are 39.8% and 52%, respectively. Such error levels are the cost to be paid for a conjoint representation of the two components, which is optimal for matrix **E** but not for the two matrices **M** and **N**. Moreover, unlike the separate analysis of the symmetric component, model (3.18) represents also the diagonal confusion percentages of matrix **M**. The approximation error decreases to 31.2% when the diagonal entries are not considered, and better levels of approximation

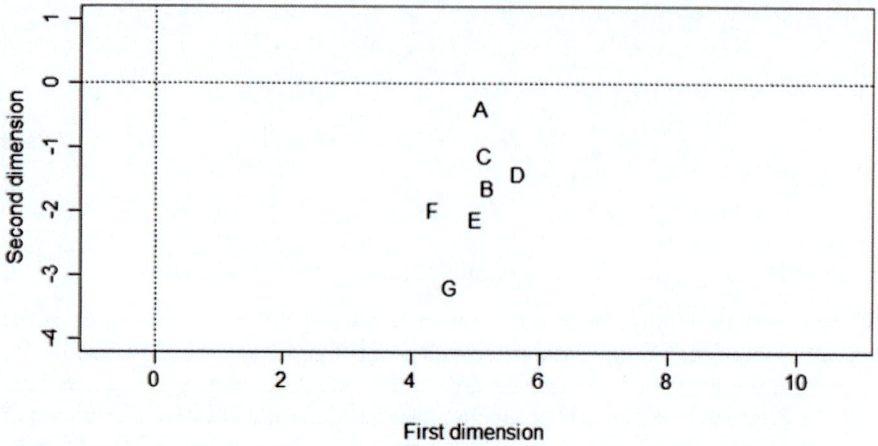

Fig. 3.10 Representation of the Escoufier–Grorud model for the seven stimuli of Table 3.1

could be obtained by setting all diagonal entries equal to the same constant value (e.g. the mean of all off-diagonal values).

In spite of the low levels of approximation, most of the information regarding the two components can be still derived from Fig. 3.10. For instance, the inner product between vectors associated with stimuli A, G is lower than the inner product between vectors C, D and between vectors B, E, reflecting the corresponding mean confusion percentages of Table 3.2. As for the separate representation in Fig. 3.2b, skew-symmetries of Table 3.3 are generally reflected in the corresponding triangle areas. For instance, the largest area is associated with the triangle with vertices at the origin and stimuli A and G, and the order in counterclockwise direction is the same of Fig. 3.2b. Moreover, diagonal entries of matrix $\mathbf{M}$ can be analysed by the lengths of the vectors, where stimulus G presents the maximum length and stimulus A the minimum.

The approximation of the proximity matrix could improve by applying the Escoufier and Grorud model in $r = 4$ dimensions (two *bimensions*). In order to simplify the graphical interpretation in four or more dimensions, a generalization of the Escoufier and Grorud model incorporating an algorithm for a type of rotation to a simple structure is proposed in Rocci and Bove [28].

3.3.2.1 DEDICOM Model and Its Constrained Versions*

The model proposed by Escoufier and Grorud [17] is a constrained version of the *DEDICOM* model proposed by Harshman [29]. This can be seen by rewriting model (3.18) in matrix notation as

$$\boldsymbol{\Omega} = \mathbf{XX}' + \mathbf{XJX}' + \mathbf{E}_1 = \mathbf{X}(\mathbf{I} + \mathbf{J})\mathbf{X}' + \mathbf{E}_1 \tag{3.20}$$

where $\mathbf{I}$ is the diagonal matrix with ones as diagonal entries (*identity matrix*) and $\mathbf{E}_1$ is the error component matrix. The relation (3.20) holds for any number of dimensions and not only for two dimensions as in (3.18), so matrix $\mathbf{X}$ can be any $n \times r$ matrix of coordinates. The *DEDICOM* model in matrix notation can be written as

$$\boldsymbol{\Omega} = \mathbf{XRX}' + \mathbf{E}_2, \tag{3.21}$$

where $\mathbf{X}$ is a $n \times r$ $(n > r)$ matrix of loadings for the n objects on the r dimensions, $\mathbf{R}$ is an $r \times r$ matrix representing the relations between the r dimensions and $\mathbf{E}_2$ is an error component matrix. Thus, model (3.20) proposed by Escoufier and Grorud [26] is obtained by setting $\mathbf{R} = (\mathbf{I} + \mathbf{J})$.

A large class of models for analysing asymmetric proximities is constrained versions of model (3.21), and examples can be found in Kiers and Takane [30], Rocci and Bove [28], Saito and Yadohisa [31] and Chino [32]. In particular, the Generalized GIPSCAL (GG) model proposed in Kiers and Takane [30] can be written, up to an additive constant,

$$\boldsymbol{\Omega} = \mathbf{X}(\mathbf{I} + \Delta)\mathbf{X}' + \mathbf{E}_3, \tag{3.22}$$

where Δ is a block diagonal matrix to be estimated having 2×2 blocks $\begin{bmatrix} 0 & \delta_s \\ -\delta_s & 0 \end{bmatrix}$ along the diagonal, and when r is odd, a zero element is in the last diagonal position. Kiers and Takane [30] remarked that GG generalizes model (3.20) (and GIPSCAL model), by allowing for a different importance for each pair of dimensions given by parameters δ_s (a minor difference in only two dimensions however). An interesting property of the GG model is that under mild conditions it allows to represent the *DEDICOM* model in diagrams, whose interpretation is similar to the Escoufier and Grorud model. A particular type of biplot representation for a reparameterization of model (3.22) was also proposed in Kiers and Takane [30], to overcome the difficulty to sum inner products and triangle areas in diagrams, when the entries of the proximity matrix $\boldsymbol{\Omega}$ have to be compared.

A general algorithm to estimate many of these constrained versions of *DEDICOM* is provided in Rocci [33]. Finally, we remark that the *DEDICOM* model is in turn a constrained version of the bilinear model presented in Chap. 2.

3.4 Overview and Strategies of Analysis

In Chaps. 2 and 3, several models and methods for graphical representation of asymmetric proximities have been presented. The focus was mainly on spatial models, but some non-spatial models will be considered in the next chapter. Some models approximate directly the entries of matrix $\boldsymbol{\Omega}$, and others the entries of the symmetric component (**M**) and/or the skew-symmetric component (**N**). Some methods are based on bilinear models (*B*), and others on distance models (*D*). Some models are constrained versions (less parameters) of other models (more parameters). In Table 3.6, an overview of some features of such models is provided (inspired by Table 23.3 in Borg and Groenen [1]), along with the indication (in the last column) of the section in this book where each model is presented. In the column labelled Graphical interpretation, the correspondence between geometrical structures and data is referred to. Each model captures a particular aspect of the asymmetry by an opportune geometrical structure, so it is worth to perform data analysis with different models to avoid to miss patterns that a single model cannot detect. When proximities are intrinsically asymmetric or their symmetric component is not interesting to analyse, models for direct representation of matrix $\boldsymbol{\Omega}$ or models for separate analysis of skew-symmetry should be preferred, respectively. Model choice could follow the hierarchy defined by the number of parameters, firstly applying parsimonious models with graphical representations easy to interpret. A separate analysis of symmetry and skew-symmetry should be preferred when an adequate level of data approximation is not obtainable by models for joint representation. Bilinear models are defined for proximities measuring pairwise similarities, and distance-based models are appropriate for proximities measuring pairwise dissimilarities. In most applications, similarity relations can be transformed into dissimilarity relations, so that the choice between bilinear or distance-based models is essentially a subjective matter.

3.5 Applications

Now, we will illustrate the capabilities of some models and methods included in Table 3.6 by using the Thomas sociomatrix presented in the first chapter (Table 1.2). The data give the liking ratings (between 0 = min. and 20 = max.) that 24 students in a school class assigned to each classmate. Diagonal entries in the table are not defined. Aim of the data analysis is to identify the pattern of affinities between the twenty-four students in the school class.

Figure 3.11 provides the frequency distribution of the off-diagonal ratings, with mean and standard deviation equal to 12.5 and 4.4, respectively. The distribution shows a moderate high dispersion (the coefficient of variation is 35.2%) and negative asymmetry (predominant high ratings), with the modal rating (the neutral judgement 10) positioned in the middle of the rating scale range.

It is interesting also to analyse and compare row and column totals of the table, because they are a first rough indication of the openness and appeal of the student with respect to the rest of the classmates, respectively. The analysis can be performed by looking at the diagram in Fig. 3.12. Each student is identified by its label as in Table 1.2. The coordinates of the student points on the horizontal and vertical axis are the corresponding row totals (total of the ratings given) and column totals (total of the ratings received), respectively. The bisector line (equation $y = x$) is also drawn in the plane to allow for a comparison between totals. For students close to the line,

Table 3.6 Overview of models for graphical representation of asymmetric proximities. A + or a − in the columns **Ω**, **M** and **N** indicates whether the model fits asymmetric proximities directly (column **Ω**), the symmetric part separately (column **M**) and the skew-symmetric part separately (column **N**). In column Model type, B stands for bilinear and D for distance. In column Parameter number, the number of dimensions r is assumed even and n is the number of rows of **Ω**

Model	Model type	Parameter number	**Ω**	**M**	**N**	Graphical interpretation	Sec
Bilinear	B	$2\,r\,n$	+	−	−	Inner products between row points and column points (*biplot type*)	2.2
Okada–Tsurumi	B	$2\,r\,n$	+	−	−	Rectangle areas determined by row and column coordinates	2.2
General GIPSCAL	B	$r\,n + r/2$	+	−	−	Symmetry by inner products, skew-symmetry by triangle areas plus orientation (a *biplot-type* representation is also possible)	3.3.2.1
Escoufier–Grorud	B	$r\,n$	+	−	−	Symmetry by inner products, skew-symmetry by triangle areas plus orientation	3.3.2
Inner product	B	$r\,n$	−	+	−	Symmetry by inner products	3.2
SVD	B	$r\,n$	−	−	+	Skew-symmetry by triangle areas plus orientation (*Gower diagram*)	3.2
Unfolding	D	$2\,r\,n$	+	−	−	Distances between row points and column points	2.3.2.1
Asymscal	D	$2\,r\,n$	+	−	−	Distances from a reference row point to the other row points	2.3.2.2
Slide vector	D	$r\,n + r$	+	−	−	Distances and projections on a reference vector	2.3.2.3
Weeks–Bentler	D	$r\,n+n+2$	+	−	−	Symmetry by distances, skew-symmetry by signed distances on line	3.2.1.2
Radius–distance	D	$r\,n + n$	−	+	+	Symmetry by distances, skew-symmetry by difference between radii	3.3.1
DistanceSkew1	D	$r\,n + n$	−	−	+	Skew-symmetry size by distances, skew-symmetry sign by radii	3.2.2

(continued)

Table 3.6 (continued)

Model	Model type	Parameter number	Ω	M	N	Graphical interpretation	Sec
DistanceSkew2	D	*r n*	–	–	+	Skew-symmetry by distances plus orientation	3.2.2
DriftVectors	D	*r n*	–	+	+	Symmetry by distances, skew-symmetry by a summary vector	3.3.2
Distance	D	*r n*	–	+	–	Symmetry by distances	3.2
1-dimens.-Skew	D	*n*	–	–	+	Skew-symmetry by signed distances on line	3.2.1

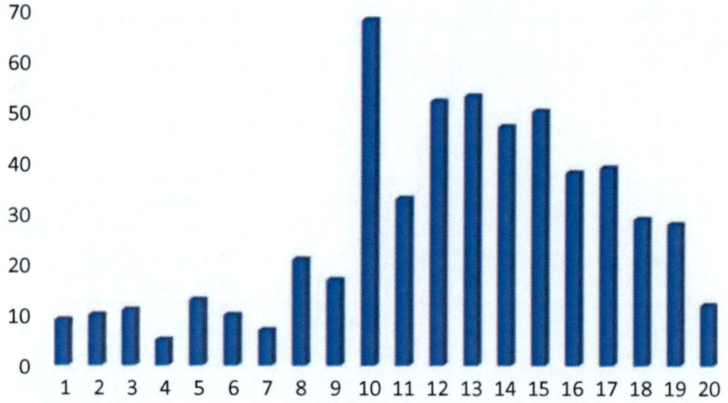

Fig. 3.11 Frequency distribution of the ratings of the Thomas sociomatrix

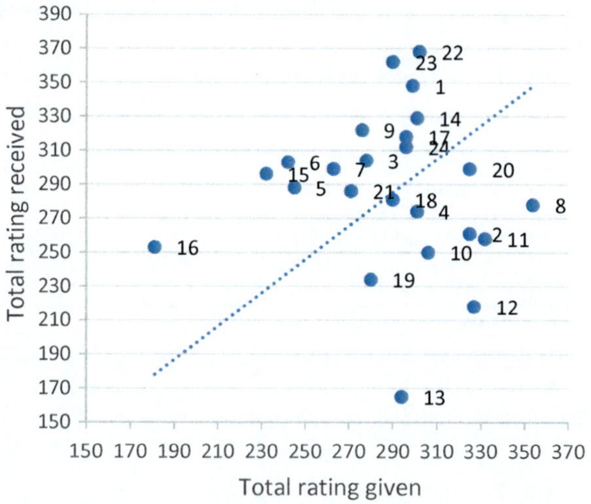

Fig. 3.12 Scatter diagram of total ratings given and received in Table 1.2

the given ratings match the received ratings (similar row and column totals). For students above the line, the received ratings exceed the given ratings (larger column than row totals), whereas for students below the line the given ratings exceed the received ratings (smaller column than row totals). For instance, students 22 and 23 present the highest totals of received ratings that means they are very appealing with respect to the rest of their classmates. They also present quite high totals of given ratings (even if lower than totals of received ratings), which means that they are open to the rest of the school class.

Therefore, students 22, 23, 1, 14 seem very well integrated with their classmates and to a lesser extent, also students 9 and 17. Student 16, in the opposite position with respect to students 22 and 23, results in a quite isolated student, having the lowest total rating given and a low total rating received. A peculiar situation characterizes student 13, who probably would like to be integrated (his total rating given is not low) but who is not appealing to the rest of the class (his total rating received is the lowest).

The pattern of relationships between the twenty-four students in the school class can be analysed in detail by models and methods included in Table 3.6 which are not influenced by the missing diagonal entries. In order to get a joint representation of symmetry and skew-symmetry, the drift vector model has been first applied. The computation of the coordinates for the representation of the symmetric component was obtained by the PROXSCAL routine in IBM-SPSS, and a linear transformation with negative slope is assumed in the minimization of the normalized raw stress.

Figure 3.13 provides the scree plot of the *Stress*-I values obtained for the first six dimensions. The shape of the line shows a first evident elbow at 2 dimensions (*Stress*-I $= 0.26$), and a second moderate elbow at 3 dimensions (*Stress*-I $= 0.19$), which suggest analysing both configurations.

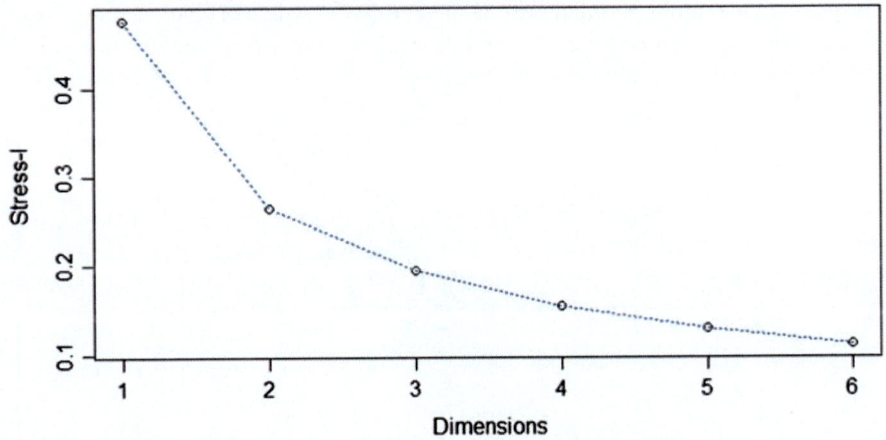

Fig. 3.13 Scree plot for the average ratings of the Thomas sociomatrix

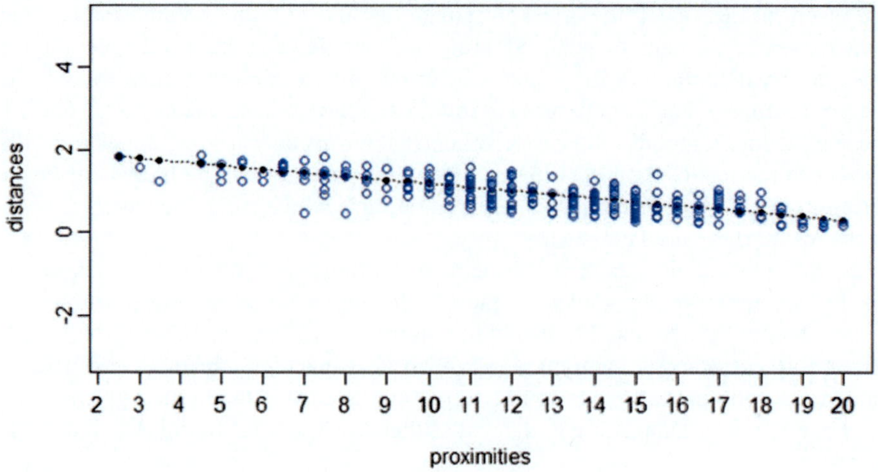

Fig. 3.14 Shepard diagram for the average ratings of Thomas sociomatrix

The Shepard diagram is provided in Fig. 3.14. The open circles represent the pairs of corresponding proximities/distances (ω_{ij}, d_{ij}) between students. Typically, the pairs of corresponding proximities and disparities $(\omega_{ij}, \hat{d}_{ij})$ (represented by the small solid points) are also displayed in the diagram. The vertical line that can be drawn from each open circle to the regression line $(\hat{d}_{ij} = -0.088\ \omega_{ij} + 2.047)$ represents the residual term $(\hat{d}_{ij} - d_{ij})$ of the point itself in the raw stress (2.6). The scatter plot provides an overall picture of the dispersion around the regression line, measured by the raw stress. The smaller the raw stress, the closer to the straight line the open circles.

Figure 3.15 exhibits the diagram of the residuals. A reasonable scatter around the regression line can be observed, and there are no outliers although some points

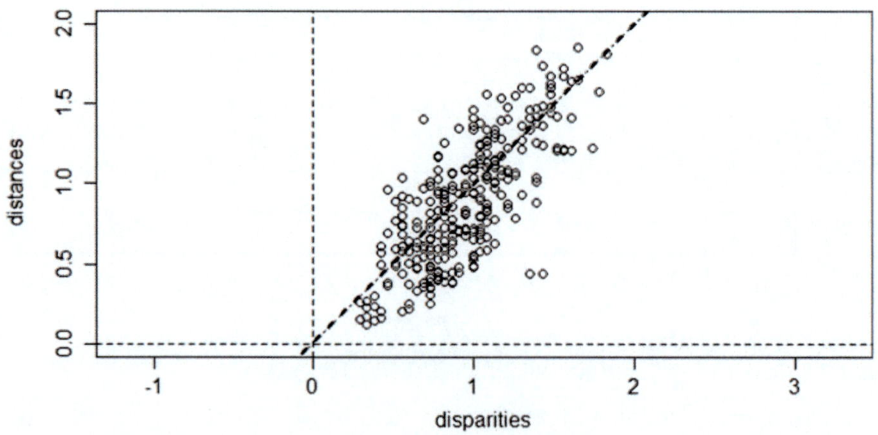

Fig. 3.15 Diagram of distances vs disparities for average ratings of the Thomas sociomatrix

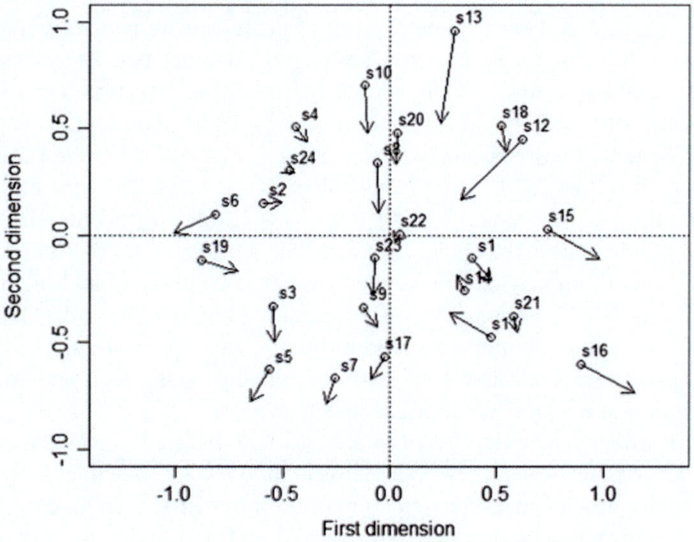

Fig. 3.16 Representation of the Thomas sociomatrix by the *drift vector* model

contribute more than others to stress. The correlation coefficient $r_{\hat{d}d}$ is 0.75, the corresponding VAF is 0.57 (57% variance accounted for), and DAF is 0.93.

The diagram of the drift vectors is presented in Fig. 3.16. Distances between points represent the symmetry, and lengths and directions of the drift vectors represent the skew-symmetry. The level of approximation of the symmetric component by the distances is not very good (*Stress*-I = 0.26), but some interesting information can be derived for symmetry and skew-symmetry. Two systematic trends can be detected by looking at the vector orientations, a group is oriented to the central area of the diagram (vectors positioned in the upper part of the plane), and another group points to the bottom of the diagram (in particular vectors in the central or lower part of the plane). This is an indication of not random asymmetry in the proximities. Many drift vectors oriented towards the central area correspond to points positioned under the line in Fig. 3.12 (students 2, 4, 8, 10, 11, 12, 13, 20), which represent students receiving total ratings lower than the total ratings they give (an indication of prevalence of positive skew-symmetries). Vectors oriented to the bottom of the diagram are mainly positioned above the line in Fig. 3.12 (students 1, 3, 5, 7, 9, 16, 17, 21, 22, 23) and identify students who received total ratings higher than the corresponding total ratings given (an indication of prevalence of negative skew-symmetries). Students 22 and 23 are positioned in the central part of the diagram, which means that their average ratings are generally high or at least not low. The skew-symmetries of student 23 should be higher than the skew-symmetries of student 22, because the drift vector of student 23 is longer than the drift vector of student 22.

Similar comments hold for students 1, 14 and 9 who are also located near the central area of the diagram. The isolation of student 16 is confirmed by his peripheral position, and the direction of the corresponding drift vector is a consequence of a prevalence of negative skew-symmetries (ratings received are generally higher than ratings given). Student 13 is also in a peripheral position, but the corresponding drift vector is directed towards the rest of the students, which is a consequence of a prevalence of positive skew-symmetries (ratings given are higher than ratings received). Clusters of close points indicate patterns of affinities between classmates. For instance, students 1, 11, 14, 21 or students 2, 4, 6, 24 are very close in the diagram and they are two examples of groups of well-related students (drift vectors are differently oriented, so they have different skew-symmetries with the classmates).

The low level of approximation of the symmetric component and the need of a detailed analysis of the pairwise skew-symmetries suggest to perform also a separate analysis of the two components (*Constantine–Gower approach*). The contributions of the symmetric and skew-symmetric components to the decomposition of the proximity matrix are 96.4% and 3.6%, respectively.

Therefore, a distance model in a three-dimensional space and an SVD model were applied to the symmetric and skew-symmetric components, respectively. Figure 3.17 shows the three-dimensional representation of the symmetric component, and since the level of approximation is improved (*Stress*-I $= 0.19$), the pattern of affinities between students is more reliable than in Fig. 3.16.

A segment perpendicular to the plane $z = 0$ identifies each student point. The two groups of students 1, 11, 14, 21 and 2, 4, 6, 24 detected in two dimensions are only

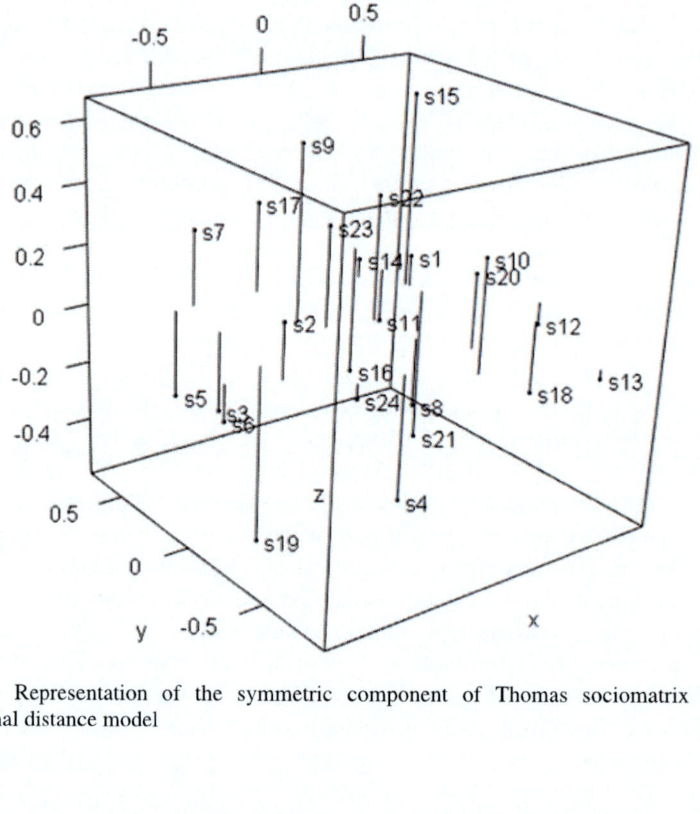

Fig. 3.17 Representation of the symmetric component of Thomas sociomatrix by three-dimensional distance model

partially confirmed in the representation, because student 21 is far from the other three students and students in the second group are all (at a different extent) distant from each other. Other groups of related students unidentifiable in two dimensions can be found in the space (e.g. 3, 5, 6 or 8, 4, 21). It is worth noting that the definition of a partition of the twenty-four students in groups of affinity is possible by using methods of cluster analysis applied to the symmetric component or methods of cluster analysis presented in Chap. 4.

A detailed analysis of the skew-symmetric component can be performed from the diagram (*Gower diagram*) in Fig. 3.18 obtained by the SVD model, by using the function *skewsymmetry* in the R package named *Asymmetry* (see the software section).

The variance accounted for is 59.4%. The analysis and comparison of the pairwise skew-symmetries between students can be performed by looking at triangle areas and orientation in the plane, and taking into account that the skew-symmetry is positive counterclockwise. For instance, since student 13 is one of the first ones in counterclockwise direction, going from student 13 to any other determines positive triangle areas: this means that the ratings given by student 13 are generally higher than the ratings he received. Such interpretations confirm what found by analysing Figs. 3.12 and 3.16. The triangle area that student 13 forms with student 6 and the

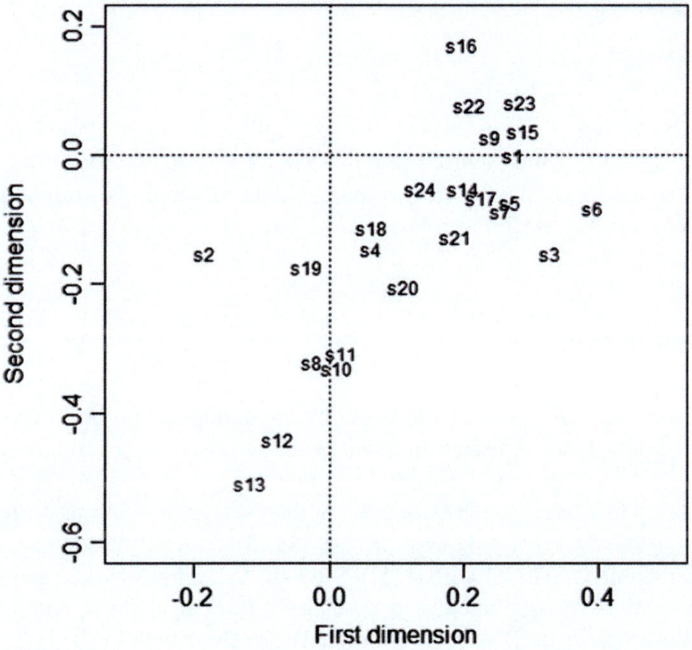

Fig. 3.18 Representation of the skew-symmetric component of Thomas sociomatrix by SVD model (*Gower diagram*)

origin is one of the largest and actually corresponds to the largest positive skew-symmetry for student 13. Student 2 and student 19 are the only exception: student 13 presents negative skew-symmetries with students 2 and 19 (Table 1.2) as correctly represented in the diagram for student 2 and at lesser extent for student 19. Similarly, students 2 and 12 have given ratings generally higher than received ratings. Student 2 is closer to the origin than students 12 and 13, so that triangle areas related to student 2 are generally smaller than triangle areas related to the other two students, which reflects the sizes of the corresponding skew-symmetry ratings of student 2. For the same reason, points close to the origin (students 4, 18 and 24) determine low skew-symmetries, and the ratings given are generally balanced with the ratings received.

Students 16, 22 and 23 are in the opposite situation with respect to points 12 and 13, because they are the last in the counterclockwise direction, with the ratings given generally lower than the ratings received.

At the end of this application, we remark that the approach presented in the analysis of the Thomas sociomatrix was mainly descriptive (see also Bove and Critchley [34]) while an analysis of the same data with a probabilistic approach by using the quasi-independence and quasi-symmetry models can be found in Caussinus and de Falguerolles [14].

3.6 Software

In this section, information concerning computer programs which allow for the application of some models and methods presented in Chap. 3 is provided. The R scripts serve to allow for an easy application of the methods, although they might not be the most efficient way for model estimations.

3.6.1 Distance Model

The distance model (3.5) can be estimated by any computer program or statistical package including routines for multidimensional scaling (for a general overview, the reader is referred to Borg and Groenen [1, Appendix A]). For the application to the matrix reported in Table 3.2, the estimation of the distance model was performed by the program PROXSCAL available in IBM-SPSS (Statistics Professional Edition or the add-on module CATEGORIES, [35]). PROXSCAL allows for the estimation of most MDS models with technically up-to-date solution algorithms, and it is one of the two MDS modules in IBM-SPSS (for details, see Borg et al. [36]). The following commands produced the output for the solution of the distance model in two dimensions for the data in Table 3.2 (including the coordinates of the points in the diagram in Fig. 3.2a).

```
DATASET ACTIVATE Dataset1.
PROXSCAL VARIABLES=A B C D E F G
   /SHAPE=BOTH
   /INITIAL=SIMPLEX
   /TRANSFORMATION=INTERVAL
   /PROXIMITIES=SIMILARITIES
   /ACCELERATION=NONE
   /CRITERIA=DIMENSIONS(2,2) MAXITER(100) DIFFSTRESS(.0001)
   MINSTRESS(.0001)
   /PRINT=COMMON DISTANCES TRANSFORMATIONS INPUT STRESS
   /PLOT=COMMON TRANSFORMATIONS(ALL) RESIDUALS(ALL).
```

As already mentioned, many other MDS programs available in statistical packages could be used, as well as the *mds* routine available in the SMACOF R package [24].

3.6.2 SVD of the Skew-Symmetric Component

The estimation of the SVD model of the skew-symmetric component of a proximity matrix and the corresponding *Gower diagram* can be computed by the function *skewsymmetry* of the Asymmetry R package written by Berrie Zielman, described at https://rdrr.io/cran/asymmetry/. As already mentioned in Sect. 2.6, besides the decomposition of the asymmetry in the symmetric and skew-symmetric components, the package provides a heat map for skew-symmetric data, the slide vector model, a scaling model with unique dimensions and the Asymscal model for asymmetric multidimensional scaling. The following R commands can be used to obtain the diagram in Fig. 3.2b.

```
library(asymmetry)
#
# Data input of the skew-symmetric component matrix
# and labels
#
input < - c(0,-4.15,-2,-0.75,-5.1,-6.8,-10.05,
        4.15,0,-0.1,-0.6,-0.95,-1.05,-4.8,
        2,0.1,0,-1.4,-0.75,-5.95,-5.8,
        0.75,0.6,1.4,0,-2.45,0.55,-10,
        5.1,0.95,0.75,2.45,0,-0.2,-1.35,
        6.8,1.05,5.95,-0.55,0.2,0,-1,
        10.05,4.8,5.8,10,1.35,1,0)
data < - matrix(input, 7,7, byrow = TRUE)
colnames(data) < - rownames(data) < - LETTERS[1:7]
#
# Computation of the SVD of the skew-symmetric component
#
Q < - skewsymmetry(data)
#
# Computation of the diagram in two dimensions (Gower diagram)
#
plot(Q, plot.plane = 1, yplus = -0.045, asp = 1)
```

3.6.3 Distance model plus radii for the skew-symmetric component

The following R program allows for the application of the two-step procedure for the representation of the skew-symmetric component of a proximity matrix by a distance model plus radii, presented in Sect. 3.2.2. The program provides as output the representation of Fig. 3.4.

```
library(smacof)
library(ggplot2)
library(ggforce)
#
# Data input of the skew-symmetric component matrix
# and labels
#
input < - c(0,-4.15,-2,-0.75,-5.1,-6.8,-10.05,
          4.15,0,-0.1,-0.6,-0.95,-1.05,-4.8,
          2,0.1,0,-1.4,-0.75,-5.95,-5.8,
          0.75,0.6,1.4,0,-2.45,0.55,-10,
          5.1,0.95,0.75,2.45,0,-0.2,-1.35,
          6.8,1.05,5.95,-0.55,0.2,0,-1,
          10.05,4.8,5.8,10,1.35,1,0)
data < - matrix(input, 7,7, byrow = TRUE)
colnames(data) < - rownames(data) < - LETTERS[1:7]
#
# MDS of the absolute values of the skew-symmetric
# component matrix data
#
MDS < - mds(delta = abs(data), ndim = 2)
#
# Computation of the radii
#
radiusConstant < - 1
linear < - rowMeans(sign(data))
linear.normalisation < - linear - linear[which.min(linear)]
#
# Normalization of the radii
#
tab < - data.frame(coord = 4.371 * MDS$conf, radius = radiusCon-
stant * linear.normalisation)
#
# Graphical representation
#
xlab < - paste("Dimension", 1, sep = " ")
ylab < - paste("Dimension", 2, sep = " ")
plot  <  -  ggplot(tab,  aes(x  =  coord.D1,  y  =  coord.D2,label
=  rownames(tab)))  +  geom_point()  +  geom_circle(aes(x0  =
tab$coord.D1,  y0  =  tab$coord.D2,  r  =  tab$radius),  inherit.aes
= FALSE) +
 geom_text(hjust = 0.5, vjust = --1, size = 5) +
xlim(min(tab[,1:2]) - max(tab$radius),max(tab[,1:2]) +
max(tab$radius)) + ylim(min(tab[,1:2]) - max(tab$radius),
```

```
max(tab[,1:2]) + max(tab$radius)) + coord_fixed() +
 theme_classic() + labs(x = xlab, y = ylab) + theme(legend.position
= "none")
plot
```

3.6.4 Joint Representation

Radius-distance model

The following R program allows for the application of the radius–distance model presented in Sect. 3.3.1 to data of Table 3.1. A configuration for the symmetric component has to be computed externally and provided as input to the program. The program provides as output the representation of Fig. 3.6.

```
#
## Radius-distance model
#
#  the  external  configuration  of  the  symmetric  compo-
nent of matrix Omega is named config
## radii are obtained by the skew-symmetric matrix N of a prox-
imity matrix Omega
#
library(ggplot2)
library(ggforce)
#
# Data input of matrix Omega
#
input < - c(84,15.1,21,20.5,5.8,11.8,0.6,
        23.4,90,11.3,20.4,22.8,14.4,15,
        25,11.5,87,24.7,18.5,7,9.4,
        22,21.6,27.5,91,15,14.8,8.4,
        16,24.7,20,19.9,92,17.2,18.8,
        25.4,16.5,18.9,13.7,17.6,90,15,
        20.7,24.6,21,28.4,21.5,17,97)
Omega < - matrix(input, 7,7, byrow = TRUE)
#
## computation of the skew-symmetric component N
#
N < - (Omega - t(Omega))/2
#
## input configuration of the symmetric component of Omega
#
config < - c(-0.701, 0.105,
          0.303,   0.240,
          -0.482,    -0.484,
          -0.190,    -0.378,
          0.596,    0.136,
          -0.157,     0.794,
          0.632,    -0.413)
coord = matrix(config,7,2,byrow = TRUE)
rownames(coord) < - c("A","B","C","D","E","F","G")
```

```
#
# Computation of the radii and normalization
#
radiusConstant = 0.067
linear < - rowMeans(N)
linear.normalisation < - linear - linear[which.min(linear)]
#
# Representation of the radius-distance model
#
tab < - data.frame(coord = coord, radius = radiusCon-
stant * linear.normalisation)
xlab < - paste("First dimension", sep = " ")
ylab < - paste("Second dimension", sep = " ")
lim.min < - max(tab[,1:2])
lim.max < - max(tab[,1:2])
radius.max < - max(tab$radius)
ggplot(tab, aes(x = coord[,1], y = coord[,2])) +
    geom_circle(aes(x0 = coord[,1], y0 = coord[,2], r =
radius), inherit.aes = FALSE) +
  geom_point() + geom_text(aes(label = rownames(tab)),hjust =
0.5, vjust = -1, size = 5) +
 xlim(min(tab[,1:2]) - radius.max,max(tab[,1:2]) + radius.max) +
ylim(min(tab[,1:2]) - radius.max,max(tab[,1:2]) + radius.max) +
 theme_bw() +
  theme(panel.grid.major = element_blank(), panel.grid.minor =
element_blank()) + labs(x = xlab, y = ylab) + theme(legend.position
= "none") + coord_fixed() +
 ggtitle("")
```

Drift Vector Method

The drift vectors can be obtained by the function *driftVectors* available in the SMACOF R package [24]. The arrows provided in Figs. 3.8 and 3.9 were computed by the following "ad hoc" *R* program adapted from a program available in the supplementary files from the website of Borg et al. [36]. A configuration for the symmetric component needs to be computed externally and provided as input to the program.

```
## Drift vectors on input configuration named config.
## drifvectors are obtained by the skew-symmetric matrix N of matrix
# Omega.
#
# Data input of matrix Omega.
#
input < - c(84,15.1,21,20.5,5.8,11.8,0.6,
         23.4,90,11.3,20.4,22.8,14.4,15,
         25,11.5,87,24.7,18.5,7,9.4,
         22,21.6,27.5,91,15,14.8,8.4,
         16,24.7,20,19.9,92,17.2,18.8,
         25.4,16.5,18.9,13.7,17.6,90,15,
         20.7,24.6,21,28.4,21.5,17,97).
Omega < - matrix(input, 7,7, byrow = TRUE)
#
## Computation of skew-symmetric component
```

```
#
N < - (Omega - t(Omega))/2
#
## input configuration of the symmetric component of Omega
#
config < - c(-0.701, 0.105,
         0.303,    0.240,
         -0.482,     -0.484,
         -0.190,     -0.378,
         0.596,     0.136,
         -0.157,     0.794,
         0.632,     -0.413)
#
# Graphical representation of the symmetric component
#
coord < - matrix(config, 7,2, byrow = TRUE)
n < - dim(coord)[1]
rownames(coord) < - c("A","B","C","D","E","F","G")
cex < - 1.3
par(cex.lab = cex, cex.axis = cex, cex.main = cex)
plot(coord, type = "p", asp = 1,xlab = "First dimension",
            ylab = "Second dimension", ylim = c(-1,1), xlim = c(-
1.5, 1.5))
abline(h = 0, v = 0, lty = "dotted")
text(coord + 0.05, labels = rownames(coord), cex = 1.5)
#
# Computation of arrows and drift vectors
#
X <- coord
for (i in 1:n) {
 sx <- 0
 sy <- 0
 x  <- X[i,1]
 y  <- X[i,2]
 for (j in 1:n) {
  c <- N[i,j]/mean(Omega);
  a <- X[j,1]
  b <- X[j,2]
  slope <- (b - y)/(a - x)
     if (c != 0) {
   angle <- atan(slope);
   length1 <- abs(c)
   x1 <- x + length1 * cos(angle)
   y1 <- y + length1 * sin(angle)
   if (c > 0){
    if(b < y){
    if (slope > 0) {
    x1 <- x + sign(-c)*length1 * cos(angle)
    y1 <- y + sign(-c)*length1 * sin(angle)}
    }}
   if (c > 0){
    if(b > y){
     if (slope < 0) {
      x1 <- x + sign(-c)*length1 * cos(angle)
```

```
          y1 <- y + sign(-c)*length1 * sin(angle)}
        }}
      if (c < 0){
         if(b > y){
          if (slope > 0) {
          x1 <- x + sign(c)*length1 * cos(angle)
          y1 <- y + sign(c)*length1 * sin(angle)}}
      }
      if (c < 0) {
        if (b < y) {
         if (slope < 0) {
          x1 <- x + sign(c)*length1 * cos(angle)
          y1 <- y + sign(c)*length1 * sin(angle)}}
      }
#
# Representation of arrows
#
     arrows(x, y, x1, y1, length=0.10, col="blue", lwd=2, lty=1)
   } ## arrows
   if (i != j) {
     sx <- sx + x1 - x
     sy <- sy + y1 - y
   } }
  ## k expansion factor for mean drift vector
  k=2
   sx <- k%*%sx/n
   sy <- k%*%sy/n
#
# Representation of drift vectors
#
  arrows(x, y, x+sx, y+sy, length=0.20, col="black", lwd=1, lty=1.5)
} ## drift vector
```

Escoufier–Grorud model

The following simple R program allows for the application of the Escoufier–Grorud model presented in Sect. 3.3.2 to data of Table 3.1. The program provides as output the representation of Fig. 3.10.

```
#
# Data input of matrix Omega
#
input <- c(84,15.1,21,20.5,5.8,11.8,0.6,
         23.4,90,11.3,20.4,22.8,14.4,15,
         25,11.5,87,24.7,18.5,7,9.4,
         22,21.6,27.5,91,15,14.8,8.4,
         16,24.7,20,19.9,92,17.2,18.8,
         25.4,16.5,18.9,13.7,17.6,90,15,
         20.7,24.6,21,28.4,21.5,17,97)
Omega <- matrix(input, 7,7, byrow = TRUE)
#
## Computation of the symmetric and skew-symmetric component of
#  matrix Omega
#
```

```
M <- (Omega + t(Omega))/2
N <- (Omega - t(Omega))/2
#
## Computation of matrix E
#
E1 <- cbind(M,-N)
E2 <- cbind(N,M)
E <- rbind(E1,E2)
#
## Computation of eigen values/vectors of matrix E
#
eigenE <- eigen(E,symmetric=TRUE)$vectors%*%diag(sqrt(eigen
                    (E,symmetric=TRUE)$values))
#
## Graphical representation
#
rowcoord=-eigenE[1:7,1:2]
rownames(rowcoord)<- c("A","B","C","D","E","F","G")
plot(rowcoord, type="n", asp=1, xlab = "First dimension",
     ylab = "Second dimension", ylim=c(-4,1),xlim=c(4, 6))
abline(h=0, v=0, lty="dotted")
text(rowcoord, labels=rownames(rowcoord))
```

References

1. Borg, I., & Groenen, P. J. F. (2005). *Modern multidimensional scaling. Theory and applications.* (2nd ed.). Springer.
2. Heiser, W. J., & Busing, F. M. T. A. (2004). Multidimensional scaling and unfolding of symmetric and asymmetric proximity relations. In D. Kaplan (Ed.), *The Sage handbook of quantitative methodology for the social sciences* (pp. 25–48). Sage Publications.
3. Constantine, A. G., & Gower, J. C. (1978). Graphical representation of asymmetric matrices. *Applied Statistics, 3*, 297–304.
4. Bowker, A. H. (1948). A test for symmetry in contingency tables. *Journal of the American Statistical Association, 43*, 572–574.
5. Hubert, L. J., & Baker, F. B. (1979). Evaluating the symmetry of a proximity matrix. *Quality and Quantity, 43*, 81–91.
6. Saburi, S., & Chino, N. (2008). A maximum likelihood method for an asymmetric MDS model. *Computational Statistics and Data Analysis, 52*, 4673–4684.
7. Zielman, B., & Heiser, W. J. (1996). Models for asymmetric proximities. *British Journal of Mathematical and Statistical Psychology, 49*, 127–146.
8. Kuriki, S. (2010). Distributions of the largest singular values of skew-symmetric random matrices and their applications to paired comparisons. *Communications in Statistics–Theory and Methods, 39*, 1522–1535
9. Critchley, F. (1988). On characterization of the inner products of the space of square matrices of given order which render orthogonal the symmetric and the skew-symmetric subspaces. *Warwick Statistics Research Report 171*. University of Warwick.
10. Carroll, J. D., & Arabie, P. (1998). Multidimensional scaling. In M. H. Birnbaum (Ed.), *Measurement, Judgment and Decision Making* (pp. 179–250). Academic Press.
11. Gower, J. C. (1977). The analysis of asymmetry and orthogonality. In J. R. Barra et al. (Eds.), *Recent developments in statistics* (pp. 109–123). North Holland.
12. Gower, J. C. (2018). Skew symmetry in retrospect. *Advances in Data Analysis and Classification, 12*, 33–41.

13. Weeks, D. G., & Bentler, P. M. (1982). Restricted multidimensional scaling models for asymmetric proximities. *Psychometrika, 47*, 201–207.
14. Caussinus, H., & de Falguerolles, A. (1987). Tableaux carrés: Modélisation et methods factorielles. *Revue De Statistique Appliquée, 35*, 35–52.
15. De Rooij, M., & Heiser, W. J. (2003). A distance representation of the quasi-symmetry model and related distance models. In H. Yanai & A. Okada (Eds.), *New Developments in Psychometrics* (pp. 487–494). Springer-Verlag.
16. De Rooij, M., & Heiser, W. J. (2005). Graphical representations and odds ratios in a distance association for the analysis of cross-classified data. *Psychometrika, 70*, 99–122.
17. De Rooij, M. (2020). Asymmetric scaling models for square contingency tables: points, circles, arrows and odds ratios. In T. Imaizumi, A. Nakayama & S. Yokoyama (Eds.), Advanced studies in behaviormetrics and data science. *Behaviormetrics: Quantitative Approaches to Human Behavior* (pp. 43–61). Springer Nature.
18. Holman, E. W. (1979). Monotonic models for asymmetric proximities. *Journal of Mathematical Psychology, 20*, 1–15.
19. Nosofsky, R. M. (1991). Stimulus bias, asymmetric similarity, and classification. *Cognitive Psychology, 23*, 94–140.
20. Brossier, G. (1982). Classification hiérarchique à partir de matrices carrées non symétriques. *Statistique et Analyse des Données, 7*, 22–40.
21. Bove, G. (2012). Exploratory approaches to seriation by asymmetric multidimensional scaling. *Behaviormetrika, 39*, 63–73.
22. Okada, A., & Imaizumi, T. (1987). Non-metric multidimensional scaling of asymmetric similarities. *Behaviormetrika, 21*, 81–96.
23. Okada, A. (1990). A generalization of asymmetric multidimensional scaling. In M. Schader & W. Gaul (Eds.), *Knowledge, data and computer-assisted decisions* (pp. 127–138). Springer.
24. Mair, P., Groenen, P. J. F., & De Leeuw, J. (2020). *More on multidimensional scaling and unfolding in R: smacof Version 2.* https://cran.r-project.org/web/packages/smacof/vignettes/smacof.pdf
25. Chino, N. (1978). A graphical technique for representing the asymmetric relationships between N objects. *Behaviormetrika, 5*, 23–40.
26. Escoufier, Y., & Grorud, A. (1980). Analyse factorielle des matrices carrées non symmétriques. In E. Diday et al. (Eds.), *Data analysis and informatics* (pp. 263–276). North Holland.
27. Chino, N. (1990). A generalized inner product model for the analysis of asymmetry. *Behaviormetrika, 17*, 25–46.
28. Rocci, R., & Bove, G. (2002). Rotation techniques in asymmetric multidimensional scaling. *Journal of Computational and Graphical Statistics, 11*, 405–419.
29. Harshman, R. A. (1978). Models for analysis of asymmetrical relationships among N objects or stimuli. *Paper presented at the First Joint Meeting of the Psychometric Society and the Society for Mathematical Psychology*. McMaster University.
30. Kiers, H. A. L., & Takane, Y. (1994). A generalization of GIPSCAL for the analysis of asymmetric data. *Journal of Classification, 11*, 79–99.
31. Saito, T., & Yadohisa, H. (2005). *Data analysis of asymmetric structures. Advanced approaches in computational statistics*. Marcel Dekker.
32. Chino, N. (2012). A brief survey of asymmetric MDS and some open problems. *Behaviormetrika, 39*, 127–165.
33. Rocci, R. (2004). A general algorithm to fit constrained DEDICOM models. *Statistical Methods & Applications, 13*, 139–150.
34. Bove, G., & Critchley, F. (1993). Metric multidimensional scaling for asymmetric proximities when the asymmetry is one-dimensional. In R. Steyer (Ed.), *Psychometric Methodology* (pp. 55–60). Gustav Fischer Verlag.
35. IBM-SPSS. (2019). *IBM-SPSS Categories 26*. IBM Corp.
36. Borg, I., Groenen, P. J. F., & Mair, P. (2018). *Applied multidimensional scaling and unfolding* (2nd ed.). Springer.

Chapter 4
Cluster Analysis for Asymmetry

4.1 Preliminaries

Cluster analysis for asymmetric data has drawn less attention than the many proposals related to symmetric proximity data and has been firstly developed by generalizing the existing classical algorithms for hierarchical cluster analysis.

Along this major approach, the asymmetric proximities are regarded as a special case of two-mode two-way data where the rows and the columns of the $(n \times n)$ proximity matrix $\boldsymbol{\Omega}$ are considered two different modes. Thus, the resulting graphical representations as hierarchical trees (generalized dendrograms) are generally complicated and difficult to analyse especially when the number of objects is large.

This difficulty in displaying the results becomes greater and greater as the size of matrix $\boldsymbol{\Omega}$ increases because the number of objects to be represented doubles. On the other hand, the need of searching for groups of objects with similar behaviour typically arises when there are many objects.

Differently, along another approach, in accordance with the original form of the (one-mode two-way) data, several clustering methodologies have been proposed to explicitly deal with the asymmetry by directly taking into account that objects in rows and columns of the proximity matrix $\boldsymbol{\Omega}$ are just the same.

The underlying idea is that groups of objects may share common behaviours not only in terms of average intensities, but also for the directions of the exchanges: some clusters of objects can mainly either originate the exchanges directed to other clusters of objects or receive from other clusters and only methods specifically designed for one-mode two-way proximity data may provide an in-depth view of the exchange dynamics.

A small real application is presented here to introduce the issue: it will be used in the following sections to illustrate the different methods. For the sake of simplicity and clarity of exposition, a subset of seven countries/languages has been selected from the European Languages data (Eurobarometer-243 survey [1]) presented in Sect. 1.4 to be examined throughout the chapter, while the complete data are fully

G. Bove et al., *Methods for the Analysis of Asymmetric Proximity Data*,
Behaviormetrics: Quantitative Approaches to Human Behavior 7,
https://doi.org/10.1007/978-981-16-3172-6_4

Table 4.1 EuroLang data: number of people (%) in European countries non-speaking foreign languages (Special Eurobarometer-243 [1])

Country	Language						
	Germ	Span	French	Ital	Dutch	Engl	Rom
DE	–	99	99	99	99	51	100
ES	99	–	95	100	100	81	100
FR	96	95	–	98	100	69	100
IT	98	99	92	–	100	76	100
NL	77	100	98	100	–	40	100
UK	95	96	82	99	100	–	100
RO	98	99	89	99	100	77	–

analysed at the end of the chapter. The following seven countries/languages have been chosen for the small data set (hereafter shortly denoted "*EuroLang*") because they can be representative of the dynamics of the multilingualism in Europe:

Countries

Germany (DE), Spain (ES), France (FR), Italy (IT), The Netherlands (NL), The United Kingdom (UK), Romania (RO).

Languages

German (Germ), Spanish (Span), French (French), Italian (Ital), Dutch (Dutch), English (Engl), Romanian (Rom).

Since all methods to be presented in this chapter are applied to proximities measuring dissimilarity relationships, they have been naturally derived by subtracting the original percentages (Table 1.4) from 100, so that the resulting percentages somehow explain the diversity between native and foreign languages. Thus, Table 4.1 reports the percentages ω_{ij} of people from country i who responded that they *do not speak* foreign language j.

Table 4.1 depicts a situation very different in the skill to master at least a foreign language and reveals the presence of lack of reciprocity in speaking, as expected. For instance, the percentage of people from The Netherlands who do not speak English is the smallest (40%), while all UK respondents (100%) state that they do not have sufficient skills to hold a conversation in Dutch. The percentages on the main diagonal are not of interest here as only foreign language knowledge is being analysed.

Table 4.1 represents an example of asymmetric proximity $\mathbf{\Omega}$ with percentages measuring a kind of dissimilarity between mother tongue (country) and foreign languages.

For any cell (i, j) in Table 4.1, the average of the percentages ω_{ij} and ω_{ji} (say m_{ij}) and the imbalance (say n_{ij}), i.e. the average departure of ω_{ij} from m_{ij}, can be computed (see Sect. 3.1), giving the symmetric and skew-symmetric matrices **M** and **N** (Tables 4.2 and 4.3), respectively.

Table 4.2 Average percentages of people non-speaking foreign languages in European countries

	Language						
Country	Germ	Span	French	Ital	Dutch	Engl	Rom
DE	–	99.0	97.5	98.5	88.0	73.0	99.0
ES	99.0	–	95.0	99.5	100.0	88.5	99.5
FR	97.5	95.0	–	95.0	99.0	75.5	94.5
IT	98.5	99.5	95.0	–	100.0	87.5	99.5
NL	88.0	100.0	99.0	100.0	–	70.0	100.0
UK	73.0	88.5	75.5	87.5	70.0	–	88.5
RO	99.0	99.5	94.5	99.5	100.0	88.5	–

Table 4.3 Imbalances of people non-speaking foreign languages in European countries

	Language						
Country	Germ	Span	French	Ital	Dutch	Engl	Rom
DE	–	0.0	1.5	0.5	11.0	−22.0	1.0
ES	0.0	–	0.0	0.5	0.0	−7.5	0.5
FR	−1.5	0.0	–	3.0	1.0	−6.5	5.5
IT	−0.5	−0.5	−3.0	–	0.0	−11.5	0.5
NL	−11.0	0.0	−1.0	0.0	–	−30.0	0.0
UK	22.0	7.5	6.5	11.5	30.0	–	11.5
RO	−1.0	−0.5	−5.5	−0.5	0.0	− 11.5	–

Generally, note that the term *imbalance* will be recurring throughout the chapter because the analysis of the departure from symmetry, in terms of magnitude and sign, is the main focus of most methods presented here and gives information on both intensities and directions of the exchanges between objects.

Specifically, Tables 4.2 and 4.3 provide a deeper insight on "who speaks (or does not speaks) what".

Different amounts of people in different countries can hold a conversation in a foreign language on average, but departures of the observed percentages from the corresponding averages reveal a structure which cannot be neglected or imputed to randomness. In some cases, deviations are small or null (e.g. 95% of people in Spain and France do not speak French and Spanish, respectively). Conversely, in some other cases there is a lack of reciprocity in the knowledge of foreign languages with different deviations in the two directions: 95% of Italians and Spaniards do not speak French on average ($m_{\text{ES,FR}} = m_{\text{IT,FR}} = 95$), but, while for native Spanish and French speakers, this is reciprocal ($n_{\text{ES,French}} = n_{\text{FR,Span}} = 0$), in a different way, French people state that they do not speak Italian more than the reverse ($n_{\text{IT,French}} = -3$, $n_{\text{FR,Ital}} = 3$).

By recalling that $\mathbf{\Omega} = \mathbf{M} + \mathbf{N}$ (refer to the decomposition into a symmetric and a skew-symmetric component in Sect. 3.1) and that such an additive decomposition also holds for the sums of squares due to the orthogonality of **M** and **N**, for the data in Table 4.1, the contributions of the symmetric matrix **M** in Table 4.2 and the skew-symmetric matrix **N** in Table 4.3 are 98.96% and 1.04%, respectively. Even though the symmetric component of the proximity is strongly dominant, the asymmetry present in the data is meaningful and not negligible.

Therefore, it is possible to investigate for the presence of groups of countries sharing common attitudes and skills in speaking the same groups of foreign languages.

Generally, this is the problem arising in cluster analysis where groups (clusters) of objects are searched for, which are similar/not different with respect to a measure of proximity (similarity or dissimilarity).

The open question is: What kind of groups are we searching for?

In order to deal with such a question, the data have been firstly clustered disregarding their asymmetries by applying the ordinary hierarchical clustering to the symmetric dissimilarity matrix **M**. Let us briefly recall how such methods work.

Hierarchical agglomerative clustering: average linkage

The most commonly used classes of clustering methods are the so-called *hierarchical agglomerative* methods which produce a series of nested or hierarchical classifications of the objects obtained by recursive merging.

Specifically, in hierarchical methods, initially, each object forms a cluster as its own (the so-called singleton), i.e. the initial partition is formed by as many clusters as the number of objects. Then, the two "closest" objects are joined to form a new cluster, and this process is continued (in a stepwise manner) for joining a cluster with another cluster until, at the last step, all objects are combined into one single group. As a consequence of the recursive merging, two objects belonging to the same cluster at a given step of the algorithm continue to belong to the same cluster in the following steps: this is the crucial property in defining a hierarchy.

In the general framework of the hierarchical methods, each method differs in how it measures the "closeness" between two clusters. In fact, while at the starting step the dissimilarities between clusters are actually those between objects, after the first merging it is necessary to define how to compute the dissimilarities between clusters, which are generally a function of the pairwise dissimilarities between objects belonging to the two merging clusters.

For example, in the popular *single linkage* method, the dissimilarity between two clusters is the minimum of the dissimilarities between all the objects in the two clusters, whereas in the *complete linkage* method it is defined as the maximum one. Again, in the *average linkage* method [2], which will be referred to in the following sections, the dissimilarity between two clusters is computed as the average of the dissimilarities between all pairs of objects, where members of a pair are in different clusters (for instance, see Hartigan [3], Landau and Ster [4], and Everitt et al. [5], for an overview of methods used to cluster data).

Briefly, starting from a square ($n \times n$) dissimilarity matrix, the algorithm consists of the following steps:

1. Find and merge the two current clusters with the smallest dissimilarity;
2. Compute the dissimilarity between the new and any other cluster and obtain a dissimilarity matrix of reduced size.

The two steps are repeated in turn until all objects are agglomerated into one single group of size n.

For the EuroLang data, the classical average linkage has been applied to the symmetric dissimilarity matrix $\mathbf{M}$ (Table 4.2), where m_{ij} indicates the average percentage of people in country i (j) who *do not speak* foreign language j (i).

At the first step, the closest countries/languages result Dutch and English because $m_{\text{Dutch,Engl}} = \min_{i<j} m_{ij} = 70, \quad (i, j = 1, \ldots, 7)$.

At the second step, the dissimilarity between the new cluster (Dutch, Engl) and any other cluster is computed; for example,

$$
\begin{aligned}
m^*_{(\text{Dutch,Engl}),\text{Germ}} &= \text{mean}\big(m_{\text{Dutch,Germ}};\ m_{\text{Engl,Germ}}\big)\\
&= \text{mean}(88;\ 73) = 80.5;
\end{aligned}
$$

$$
\begin{aligned}
m^*_{(\text{Dutch,Engl}),\text{Span}} &= \text{mean}\big(m_{\text{Dutch,Span}};\ m_{\text{Engl,Span}}\big)\\
&= \text{mean}(100;\ 88.5) = 94.25;
\end{aligned}
$$

and so on for all the pairs of clusters, so that the new dissimilarity matrix $\mathbf{M}^*$ is of size $(n-1) = 6$ and replaces $\mathbf{M}$ to be used in Step 1.

Returning to Step 1, the pair of closest clusters to be merged is identified: $m_{(\text{Dutch,Engl}),\text{Germ}} = \min_{i<j} m_{ij} = 80.5, \quad (i, j = 1, \ldots, 6)$.

Again the new dissimilarities between the new cluster (Dutch, Engl, Germ) and any other cluster are computed as follows:

$$
\begin{aligned}
m^*_{(\text{Dutch,Engl,Germ}),\text{Span}} &= \text{mean}\big(m_{\text{Dutch,Span}};\ m_{\text{Engl,Span}};\ m_{\text{Germ,Span}}\big)\\
&= \text{mean}(100;\ 88.5;\ 99) = 95.83;
\end{aligned}
$$

$$
\begin{aligned}
m^*_{(\text{Dutch,Engl,Germ}),\text{French}} &= \text{mean}\big(m_{\text{Dutch,French}};\ m_{\text{Engl,French}};\ m_{\text{Germ,French}}\big)\\
&= \text{mean}(99;\ 75.5;\ 97.5) = 90.67;
\end{aligned}
$$

and so on for any pair of clusters, until at the very final step the last two remaining clusters are merged to form only one cluster including all languages.

The resulting hierarchical clustering is then displayed pictorially as a hierarchical tree referred to as *dendrogram* (Fig. 4.1), where the different heights represent the merging levels and the sequence of the agglomeration process.

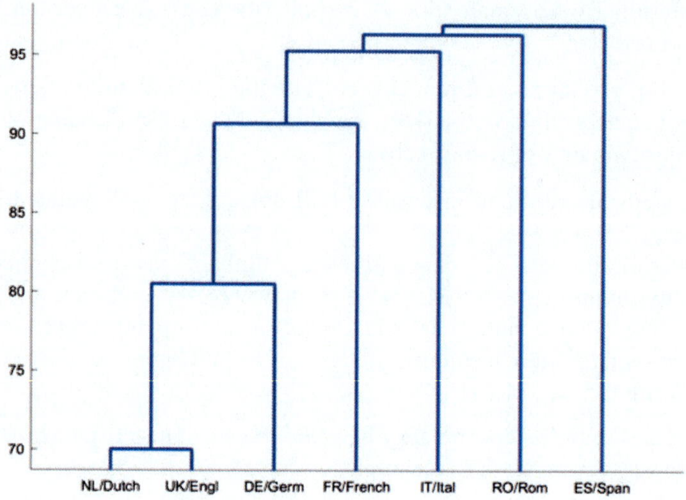

Fig. 4.1 Dendrogram from average linkage method for the symmetric part (**M**) of the EuroLang data

The dendrogram in Fig. 4.1 shows the similar countries/languages in terms of average percentages of non-spoken languages, where, for instance, NL/Dutch and UK/Engl result the most similar but where any information on the directions of "who speaks what" is lost. Again, ES/Span results to be the country/language less similar to the others, but no details on whether Spaniards master (do not master) foreign languages or whether Spanish is the dominant (most spoken) or the dominated (least spoken) language.

Thus, an additional question which remains unanswered by analysing only the average percentages could be: *Which* countries better speak (do not speak) *which* languages? And what about the presence of dominant groups of countries and languages?

Figure 4.2 displays the heat map (where the intensity of the colour of each cell is proportional to the cell value) of matrix **M** in Table 4.2 after reordering rows and columns according to the hierarchical clustering of Fig. 4.1 and shows the presence of patterns of similarity between contiguous cells which reflect what emerged by looking at the dendrogram in Fig. 4.1: close cells with similar shades present similar average percentages. When the same reordering of rows and columns is applied to matrix **N** of Table 4.3, the pattern is less evident but highlights some discordances (Fig. 4.3). Most of the cells with negative values are located in the lower triangle of the map, with the exception of the cell (NL, Engl), (NL, Germ), (NL, French) and (RO, Span). Moreover, some cells are not located near cells with similar values (i.e. $n_{\text{FR,Rom}} = 5.5$) which highlights how clustering patterns built from **M** (average percentages) are different from the corresponding clustering patterns in **N** (deviations from the average percentages).

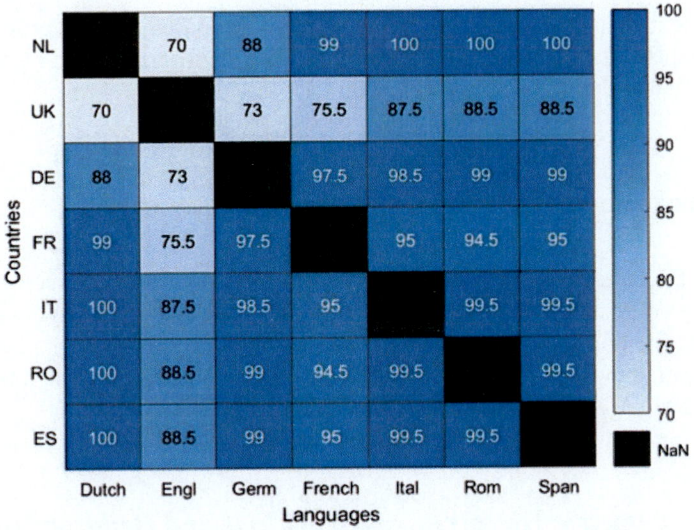

Fig. 4.2 Reordered symmetric matrix **M** of Table 4.2

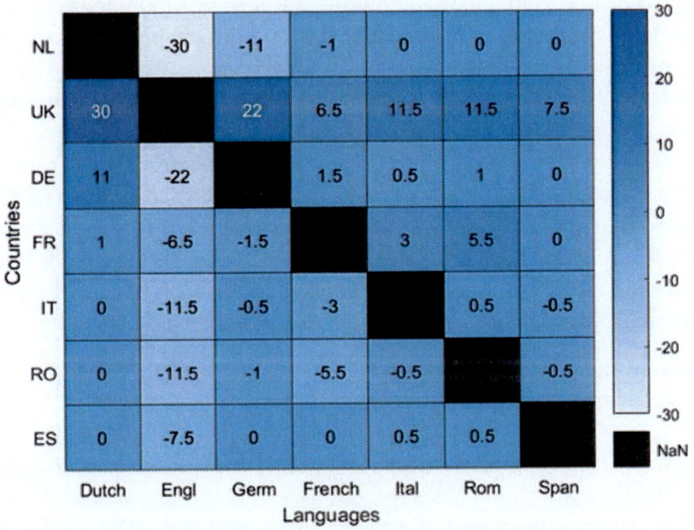

Fig. 4.3 Reordered skew-symmetric matrix **N** of Table 4.3

Generally speaking, cluster analysis methodologies designed to take explicitly into account the asymmetries and/or the skew-symmetric matrix **N** may help to find dominant (groups of) objects and to search for groups of objects having similar behaviours in terms of both average amounts and directions of the exchanges.

In the next sections, some methods specifically proposed for clustering objects from asymmetric proximities will be presented, distinguishing the two main classes of cluster analysis methods: hierarchical and non-hierarchical.

4.2 Hierarchical Methods

Following the same lines of the standard hierarchical methods for symmetric dissimilarities, several extensions have been proposed to incorporate the asymmetry into the existing methods. They are generally based on recursive algorithms where the step of updating the dissimilarities between clustered objects is properly modified to accommodate the asymmetric nature of the data.

4.2.1 Asymmetric Hierarchical Algorithms and Dendrograms

The first proposals for clustering asymmetric data date back to seventies and eighties and, generally speaking, are positioned in the framework of hierarchical clustering.

The very first asymmetric hierarchical methods, with the pioneer work by Hubert [6] and its extensions by Fujiwara [7] and Okada and Iwamoto [8], were naturally presented as extensions of the standard well-known hierarchical agglomerative algorithms of the single, complete and average linkage.

Without loss of generality, the methods are presented here starting from an asymmetric proximity matrix $\mathbf{\Omega}$ which measures dissimilarity relationships, but they can be reformulated for similarity data, straightforwardly.

Similarly to the standard hierarchical methods, using the observed asymmetric dissimilarities ω_{ij} and starting from the partition where each object forms a cluster, such algorithms consist of two different steps which alternate in turn:

1. selection of the clusters to be merged;
2. updating the dissimilarities between clusters.

which are modified properly to take into account the nature of the one-mode asymmetric data.

Briefly, the general algorithm can be sketched as follows.

Step 1 (*merging*)
Select clusters $\mathcal{C}_I$ and $\mathcal{C}_J$ to be merged forming a new cluster $\mathcal{C}_{IJ}$ ($\mathcal{C}_{IJ} = \mathcal{C}_I \cup \mathcal{C}_J$), such that, for any pair of current clusters $\mathcal{C}_P, \mathcal{C}_Q$,

$$\omega_{\mathcal{C}_I\mathcal{C}_J} = \min\big(g\big(\omega_{\mathcal{C}_P\mathcal{C}_Q}, \omega_{\mathcal{C}_Q\mathcal{C}_P}\big)\big),$$

where g indicates a combining criterion (for instance, max, min, mean) and $\omega_{C_P C_Q}$ is the dissimilarity between current clusters C_P and C_Q.

Step 2 (*updating dissimilarities between clusters*)

$$\omega^*_{C_{IJ},C_K} = f\left(\omega_{C_I C_K}, \omega_{C_J C_K}\right) \text{ and } \omega^*_{C_K,C_{IJ}} = f\left(\omega_{C_K C_I}, \omega_{C_K C_J}\right),$$
$$(\text{for any } C_K \neq C_I, C_J),$$

where f is a function of the dissimilarities to be chosen.

Steps 1 and 2 are repeated in turn until all objects form one cluster.

As special cases, by setting $g = \max$ and either $f = \min$ or $f = \max$, the algorithm without symmetrization by Fujiwara [7] follows, while the algorithm by Okada and Iwamoto [8] can be obtained when $g = \max$ and $f = \text{mean}$.

Moreover, when a preliminary STEP 0 is added where $\mathbf{\Omega}$ is symmetrized by taking either $\overline{\omega}_{ij} = \overline{\omega}_{ji} = \max\left(\omega_{ij}, \omega_{ji}\right)$ or $\overline{\omega}_{ij} = \overline{\omega}_{ji} = \min\left(\omega_{ij}, \omega_{ji}\right)$, the pioneer algorithm by Hubert [6] derives, by running iteratively Steps 1 and 2 on $\overline{\mathbf{\Omega}}$ and setting either $f = \min$ or $f = \max$.

Generalized agglomerative hierarchical method

In the framework of standard hierarchical agglomerative methods for symmetric dissimilarity data, Lance and Williams [9, 10] provided an updating formula for computing dissimilarities between clustered objects (Step 2 of the algorithm) depending on several parameters. Thus, the Lance and Williams updating formula (L&W) provides a unified framework which allows to implement a flexible algorithm including the popular standard agglomerative hierarchical algorithms as special cases (i.e. single, average, complete linkage, Ward's method).

In the context of the asymmetric hierarchical clustering, Yadohisa [11] proposed a generalized L&W incorporating the asymmetry into the formulation which gives rise to generalized expressions for updating the dissimilarities between clusters $\omega^*_{C_{IJ},C_K}$ and $\omega^*_{C_K,C_{IJ}}$ in Step 2.

Specifically, after the merging step, in computing the dissimilarity between the new cluster C_{IJ} and any other cluster C_K, Yadohisa [11] added parameters controlling the asymmetric dissimilarities due to the merged clusters C_I and C_J, so that Step 2 of the algorithm is integrated as follows:

$$\omega^*_{C_{IJ},C_K} = f_1\left(\omega_{C_I C_K}, \omega_{C_K C_I}, \omega_{C_J C_K}, \omega_{C_K C_J}, \omega_{C_I C_J}, \omega_{C_J C_I}\right) \text{ and}$$

$$\omega^*_{C_K,C_{IJ}} = f_2\left(\omega_{C_I C_K}, \omega_{C_K C_I}, \omega_{C_J C_K}, \omega_{C_K C_J}, \omega_{C_I C_J}, \omega_{C_J C_I}\right), (\text{for any } C_K \neq C_I, C_J),$$

where f_1 and f_2 are generally two different functions of the dissimilarities.

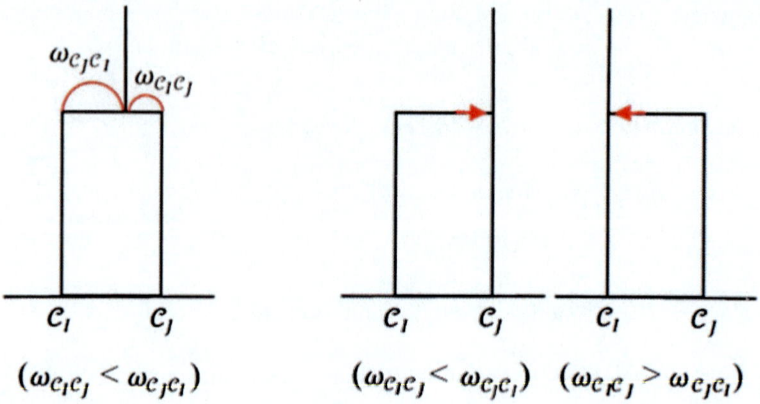

Fig. 4.4 Extended dendrograms: Okada and Iwamoto [8] (on the left), Okada and Iwamoto [13] (on the right)

Its further extension by Takeuchi et al. [12] allows to get a very general updating formula for defining a common framework including many algorithms specifically designed for asymmetric data (see Table 1 in Takeuchi et al. [12]) in addition to the well-known hierarchical algorithms for the symmetric case as well.

Extended dendrogram

Similarly to the tree graphical representations used to display the resulting hierarchies from the symmetric agglomerative algorithms, the proposals of the asymmetric hierarchical methods have been accompanied by extended dendrograms capable of visualizing the merging direction.

For instance, Okada and Iwamoto [8] display the asymmetry by representing segments proportional to $\omega_{C_I C_J}$ and $\omega_{C_J C_I}$ (derived from Step 1) as depicted in Fig. 4.4 (on the left), where $\omega_{C_I C_J}$ is the dissimilarity from objects in cluster C_I to objects in cluster C_J and $\omega_{C_J C_I}$ is the reverse one. Alternatively, Okada and Iwamoto [13] represent the asymmetry by adding arrows to the standard dendrogram as displayed in Fig. 4.4 (on the right) to represent the direction of the merging: when $\omega_{C_I C_J} > \omega_{C_J C_I}$, then cluster C_J is absorbed by cluster C_I, while when $\omega_{C_I C_J} < \omega_{C_J C_I}$, the reverse follows.

In a different fashion, Yadohisa [11] proposed an extended dendrogram which at each merging level (Step 1), in addition, includes a combined measure of $\omega_{C_I C_J}$ and $\omega_{C_J C_I}$ (for instance, min, max, mean) to be compared to $\omega_{C_I C_J}$ and $\omega_{C_J C_I}$. Such a combined measure v_{IJ} is more generally defined by Takeuchi et al. [12] as a function $v_{IJ} = g\left(\omega_{C_I C_J}, \omega_{C_J C_I}\right)$ to be computed in Step 1 of the general algorithm and represents the level at which clusters C_I and C_J merge forming cluster C_{IJ} (see Saito and Yadohisa [14], Chap. 5, for further details).

In comparison with the customary dendrograms, since such extended dendrograms incorporate three measures $\left(\omega_{C_I C_J}, \omega_{C_J C_I}, v_{IJ}\right)$ to be displayed at any merging level, they can be difficult to interpret as the number of objects increases.

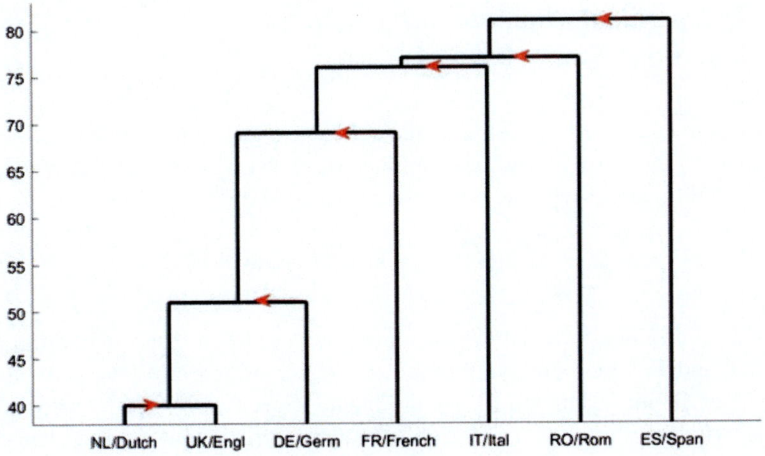

Fig. 4.5 Extended dendrogram for EuroLang data

Application of the asymmetric hierarchical algorithm to EuroLang data

The extended dendrogram in Fig. 4.5 displays the results for the EuroLang data obtained from the asymmetric hierarchical algorithm described above, setting $g =$ min in Step 1 and $f =$ min in Step 2.

The levels at which clusters merge are visualized by the different heights and represent the percentages of people from (a cluster of) countries who do not speak (a cluster of) foreign languages.

When compared with the dendrogram in Fig. 4.1 built from the average percentages regardless of the asymmetry, we may note that shapes are similar, but here the directions of the merging are also depicted: for any cluster the arrow points towards the similar cluster from which it is absorbed. Firstly, at the lowest level the arrow directed from NL to English reveals that NL presents the smallest percentage of people non-speaking English as a foreign language, while UK people do not speak Dutch at all ($\omega_{\text{NL,Engl}} < \omega_{\text{UK,Dutch}}$). At the second level, the arrow from DE points to cluster (Dutch, Engl), which indicates that DE/Germ is absorbed from the existing cluster because $\omega_{(\text{NL + UK}),\text{Germ}} > \omega_{\text{DE},(\text{Dutch + Engl})}$. Generally, the clustering process shows that in all countries the imbalance is in favour of English or the cluster it belongs to, which qualifies English as the dominant (most spoken) language. Finally, at the top of the dendrogram, SP/Span is the last to be absorbed, because Spaniards generally have the largest percentages of non-speaking foreign languages $\omega_{(\text{All Countries}),\text{Span}} > \omega_{\text{ES},(\text{All languages})}$.

4.2.2 *Independent Modelling of Symmetry and Skew-Symmetry*

The advantage of the orthogonality of the symmetric and skew-symmetric components (Chap. 3) of matrix $\mathbf{\Omega}$ has been exploited in clustering context by Brossier [15] who proposed to model separately the two matrices $\mathbf{M}$ and $\mathbf{N}$

$$\mathbf{\Omega} = \mathbf{M} + \mathbf{N} = \mathbf{A} + \mathbf{E} = (\mathbf{T} + \mathbf{K}) + \mathbf{E}, \tag{4.1}$$

where the approximation asymmetric matrix $\mathbf{A}$ is split into the two orthogonal matrices $\mathbf{T}$ and $\mathbf{K}$ which are symmetric and skew-symmetric, respectively, and $\mathbf{E}$ is the part of $\mathbf{\Omega}$ not accounted for by the model. Specifically, $\mathbf{T}$ is constrained to be ultrametric (see Sect. 1.2), i.e. $\mathbf{T}$ fulfils the ultrametric inequality and identifies a hierarchy obtained from a hierarchical tree representation (ultrametric tree or dendrogram) of $\mathbf{M}$. Moreover, $\mathbf{K}$ is the skew-symmetric approximation of rank 2 of matrix $\mathbf{N}$ (see Chap. 3)

$$\mathbf{K} = \mathbf{b}\mathbf{1}_n' - \mathbf{1}_n\mathbf{b}', \tag{4.2}$$

where $\mathbf{b}$ is a column vector of size n to be estimated and $\mathbf{1}_n$ denotes a column vector of n ones.

This means that for any pair of objects (i, j), the skew-symmetries are expressed as differences between two weights

$$k_{ij} = b_i - b_j, (i, j = 1, \ldots, n),$$

and according to (4.1) the observed dissimilarities are modelled as

$$\omega_{ij} = m_{ij} + n_{ij} = a_{ij} + \varepsilon_{ij} = t_{ij} + \left(b_i - b_j\right) + \varepsilon_{ij}, (i, j = 1, \ldots, n).$$

For identifiability reasons, vector $\mathbf{b}$ is constrained to be centred to zero mean.

Due to the independence of $\mathbf{M}$ and $\mathbf{N}$, model (4.1) can be estimated in the least-squares sense

$$LS(\mathbf{\Omega}, \mathbf{A}) = \|\mathbf{\Omega} - \mathbf{A}\|^2 = \|\mathbf{M} - \mathbf{T}\|^2 + \|\mathbf{N} - \mathbf{K}\|^2, \tag{4.3}$$

by independently minimizing the two right-hand terms subject to the ultrametric constraints on $\mathbf{T}$.

The approximation problem for the symmetric part can be easily solved by applying the standard average linkage algorithm to the symmetric matrix $\mathbf{M}$ and deriving the corresponding hierarchy (and ultrametric matrix $\widehat{\mathbf{T}}$). The use of the average linkage algorithm is motivated because it is expected to provide the best approximation of **M** in the least-squares sense, and the result is interpreted as an average level of exchange.

The least-squares estimate of parameter b_i is given by

$$\hat{b}_i = \frac{1}{n}\sum_{j=1}^{n} n_{ij}, \; (i = 1, \ldots, n), \tag{4.4}$$

as for the one-dimensional model for skew-symmetry in Sect. 3.2.1.

In order to provide a graphical representation which can help the analysis and interpretation, Brossier proposed to display the results by a simultaneous representation of the symmetric and skew-symmetric parts. This is achieved by an "additive tree" where the standard average linkage dendrogram is modified by "adding" (increasing or reducing) the lengths of the tree branches at each terminal node i of a quantity equal to $\hat{b}_i$.

Application of the Brossier method to EuroLang data

Figure 4.6 displays the results from the method by Brossier applied to the EuroLang data of Table 4.1. The dendrogram from the average linkage (Fig. 4.1), which represents the approximation $\widehat{\mathbf{T}}$ of the symmetric component **M**, has been added seven branches to the terminal nodes of lengths $\hat{b}_i (i = 1, \ldots, 7)$.

	DE Germ	ES Span	FR French	IT Ital	NL Dutch	UK Engl	RO Rom
$\hat{b}_i$	−1.14	−0.93	0.21	−2.14	−6.00	12.71	−2.71

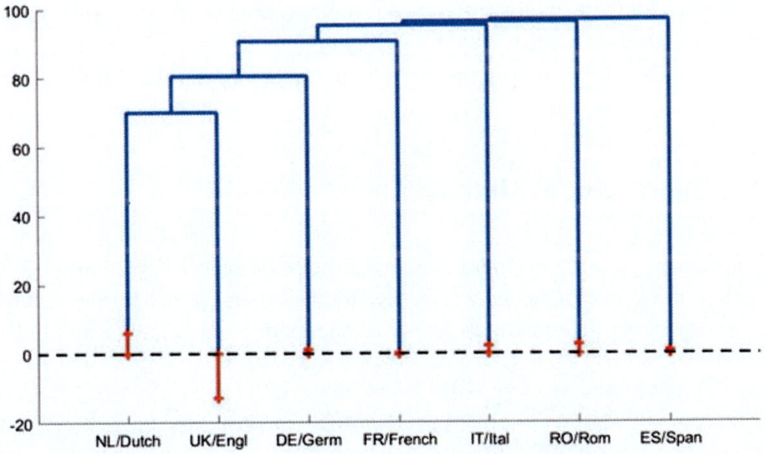

Fig. 4.6 Additive tree from the method by Brossier for EuroLang data

Table 4.4 Estimated ultrametric matrix $\widehat{\mathbf{T}}$ from average linkage for EuroLang data

	Language						
Country	Germ	Span	French	Ital	Dutch	Engl	Rom
DE	0	96.92	90.67	95.25	80.00	80.00	96.30
ES	96.92	0	96.92	96.92	96.92	96.92	96.92
FR	90.67	96.92	0	95.25	90.67	90.67	96.30
IT	95.25	96.92	95.25	0	95.25	95.25	96.30
NL	80.00	96.92	90.67	95.25	0	70.00	96.30
UK	80.00	96.92	90.67	95.25	70.00	0	96.30
RO	96.30	96.92	96.30	96.30	96.30	96.30	0

The estimated ultrametric $\widehat{\mathbf{T}}$ is reported in Table 4.4: For any pair of objects (i, j), the ultrametric distance $\hat{t}_{ij}$ indicates the level at which i and j merge to the same cluster and corresponds to the horizontal line joining i and j in the tree (Fig. 4.6).

By comparing Figs. 4.1 and 4.6, it emerges the additional information on the linguistic flow directions: a positive length indicates a surplus, whereas a negative value indicates a deficit. Specifically, NL and UK present similar percentages on average, but English is the dominant (most spoken) language, while NL is the country where people speak foreign languages the most. The greater the length of the branch, the higher the degree of asymmetry of the country/language.

The model fits well, because the proportion of the variance of the percentages of people not speaking a foreign language accounted for is equal to 99.36% (i.e. $\left(\widehat{\mathbf{T}}+\widehat{\mathbf{K}}\right)$ explains 99.36% of $\mathbf{\Omega}$, where $\widehat{\mathbf{K}}$ is estimated from (4.2) by using the estimated $\widehat{\mathbf{b}}$).

Note that the information from Fig. 4.6 is consistent with Fig. 4.5, but it is also worth noting that the asymmetric clustering methods in Sect. 4.2.1 work directly on the asymmetries $\mathbf{\Omega}$. Conversely, the Brossier method searches for clusters only from the symmetric part (**M**), regardless of the directions of the flows whose information is superimposed to the dendrogram, but not actually involved in the clustering process.

4.3 Non-hierarchical Methods

Broadly speaking, non-hierarchical clustering methods search for a partition of the objects in a number of clusters generally provided by the user, so that partitions with different numbers of clusters are obtained in independent runs and are not linked by nesting constraints.

Even if non-hierarchical methods are the most used for two-mode two-way data (such as the popular k-means algorithm), less attention has been paid to the development of methods for symmetric and, even less, for asymmetric proximity data. In the following sections, models specifically designed for one-mode two-way data are presented. Only a brief mention will be made of the kind of methods which, originally proposed for two-mode two-way data, can be applied to one-mode asymmetric proximity data as well.

4.3.1 Joint Modelling Symmetry and Skew-Symmetry

Following an approach where symmetries and skew-symmetries are jointly involved in the clustering process, Vicari [16] introduced a non-hierarchical model which relies on the decomposition of the dissimilarity matrix $\mathbf{\Omega}$ into symmetric and skew-symmetric effects both decomposed into within and between cluster effects. Differently from Brossier [15], the idea here is that not only the average amounts in matrix $\mathbf{M}$, but also the imbalances in $\mathbf{N}$ should contribute to build the clusters of objects. Therefore, the two clustering structures modelling the independent components are assumed to share the same unique partition where objects belonging to the same clusters:

(A) share the same behaviours in terms of exchanges directed to the other clusters, i.e. some clusters are mainly origins towards (destinations from) some other clusters;
(B) form "closed systems" in the sense that internal exchanges do not affect the exchanges with the other clusters.

Actually, the very general model by Vicari [16] may assume different partitions for the symmetries and skew-symmetries due to their independence, but the restricted case with a unique partition is useful to identify objects similar for both average amounts and imbalances, i.e. same intensities and directions of the exchanges.

For example, clusters may identify weak versus strong brands, exporting versus importing countries, groups of social actors having the same interactions with other social groups: in such a respect, clusters are more useful and meaningful when jointly subsume the information on amounts and directions of the exchanges.

The model

The dissimilarity matrix $\mathbf{\Omega} = \mathbf{M}+\mathbf{N}$ is modelled by defining two clustering structures for $\mathbf{M}$ and $\mathbf{N}$ depending on a unique common partition into K clusters. It is worth to remind that any partition of a set of n objects in K non-overlapping clusters $\{C_1, \dots C_j, \dots, C_K\}$ can be uniquely identified by a binary membership matrix $\mathbf{U} = [u_{ij}]$, where $u_{ij} = 1$ if object i belongs to cluster C_j, and $u_{ij} = 0$, otherwise. Moreover, object i is assumed to belong to one and only one cluster: $\sum_{j=1}^{K} u_{ij} = 1, (i = 1, \dots, n)$.

Furthermore, let us recall that for any partition, the decomposition of the data into within and between components generally holds and such components are independent.

Symmetries **M** and skew-symmetries **N** are modelled by clustering structures **S** and **K**, respectively, further decomposed into their within and between components

$$\mathbf{M} = \mathbf{S} + \mathbf{E}_S = \mathbf{S}^W(\mathbf{U}) + \mathbf{S}^B(\mathbf{U}) + \mathbf{E}_S$$

$$\mathbf{N} = \mathbf{K} + \mathbf{E}_K = \mathbf{K}^W(\mathbf{U}) + \mathbf{K}^B(\mathbf{U}) + \mathbf{E}_K,$$

where $\mathbf{S}^W(\mathbf{U})$ and $\mathbf{S}^B(\mathbf{U})$ denote the within and between components of **S** due to partition **U**, while $\mathbf{K}^W(\mathbf{U})$ and $\mathbf{K}^B(\mathbf{U})$ are the corresponding parts of **K**; $\mathbf{E}_S$ and $\mathbf{E}_K$ are the parts of **M** and **N** not accounted for by the models, respectively.

Approximation of the symmetric component

In Vicari [16], the symmetric within and between clustering structures $\mathbf{S}^W(\mathbf{U})$ and $\mathbf{S}^B(\mathbf{U})$ is assumed to be the partitioning models proposed by Vichi [17]

$$\mathbf{S}^W(\mathbf{U}) = \mathbf{U}\mathbf{D}^W\mathbf{U}' - \mathrm{diag}\big(\mathbf{U}\mathbf{D}^W\mathbf{U}'\big) \tag{4.5}$$

$$\mathbf{S}^B(\mathbf{U}) = \mathbf{U}\mathbf{D}^B\mathbf{U}', \tag{4.6}$$

where $\mathbf{D}^W = \left[d_{jj}^W\right]$ is a $(K \times K)$ diagonal matrix where entry d_{jj}^W represents the heterogeneity *within* cluster j and $\mathbf{D}^B = \left[d_{jl}^B\right]$ is a $(K \times K)$ symmetric matrix where d_{jl}^B measures the separation *between* clusters j and l. Additional ultrametric constraints on the triplets of objects may be also required for matrix $\mathbf{D}^B$, so that a "*hierarchical partitioning model*" is defined, (i.e. a partition with hierarchical structure between clusters), and $\mathbf{D}^B$ results to be a K-ultrametric matrix with only K different levels [17].

From (4.5) and (4.6), the approximation matrix **S** becomes

$$\mathbf{S} = \mathbf{S}^W(\mathbf{U}) + \mathbf{S}^B(\mathbf{U}) = \mathbf{U}\mathbf{D}^W\mathbf{U}' - \mathrm{diag}\big(\mathbf{U}\mathbf{D}^W\mathbf{U}'\big) + \mathbf{U}\mathbf{D}^B\mathbf{U}' \tag{4.7}$$

subject to

- **U** binary membership matrix
- $\mathbf{D}^B$ ultrametric.

Approximation of the skew-symmetric component

An approximation matrix **K** of the skew-symmetric component **N** can be expressed in a canonical form as a sum of a number of skew-symmetric matrices of rank

2. Therefore, by considering a partition into K clusters and the orthogonal column vectors $\mathbf{u}_j (j = 1, \ldots, K)$ of its membership matrix $\mathbf{U}$, Vicari [16] proposed to model the *within* and *between* cluster effects as follows:

$$\mathbf{K}^W(\mathbf{U}) = \sum_{j=1}^{K} (\mathbf{w}_j \mathbf{u}_j' - \mathbf{u}_j \mathbf{w}_j') = (\mathbf{W}\mathbf{U}' - \mathbf{U}\mathbf{W}') \tag{4.8}$$

$$\mathbf{K}^B(\mathbf{U}) = \sum_{j=1}^{K} \left(\mathbf{b}_j (\mathbf{1}_n - \mathbf{u}_j)' - (\mathbf{1}_n - \mathbf{u}_j)\mathbf{b}_j'\right) = (\mathbf{B}(\mathbf{1}_{nK} - \mathbf{U})' - (\mathbf{1}_{nK} - \mathbf{U})\mathbf{B}'), \tag{4.9}$$

where $\mathbf{w}_j$ and $\mathbf{b}_j$ are the orthogonal columns of the $(n \times K)$ matrices $\mathbf{W} = [\mathbf{w}_1, \ldots, \mathbf{w}_j, \ldots \mathbf{w}_K]$ and $\mathbf{B} = [\mathbf{b}_1, \ldots, \mathbf{b}_j, \ldots \mathbf{b}_K]$, respectively, both having the same structure of $\mathbf{U}$, i.e. non-null and zero entries in the same positions of $\mathbf{U}$, and $\mathbf{1}_{nK}$ denotes the $(n \times K)$ matrix of ones.

Actually, $\mathbf{W}$ and $\mathbf{B}$ are matrices which allow to express the imbalances within and between clusters as differences of weights, and, for identifiability reasons, they are constrained to have column sums equal to zero ($\mathbf{1}_n'\mathbf{W} = \mathbf{0}_n$) and to sum to zero ($\mathbf{1}_n'\mathbf{B}\mathbf{1}_K = 0$), respectively.

Specifically, any entry in (4.8) is given by

$$k_{il}^W = \sum_{j=1}^{K} (w_{ij} - w_{lj}) u_{ij} u_{lj}, \quad (i, l = 1, \ldots, n),$$

where any term measures the imbalance between objects (i, l) belonging to the same cluster j. Similarly, any entry in (4.9) is given by

$$k_{il}^B = \sum_{j=1}^{K} [b_{ij}(1 - u_{lj}) - (1 - u_{ij}) b_{lj}], \quad (i, l = 1, \ldots, n),$$

and represents the imbalance between objects (i, l) belonging to different clusters.

Note that, since the columns of $\mathbf{W}$ sum to zero, clusters form systems of internal exchanges where the within imbalances from objects in one cluster towards objects in the same cluster are null on average. Moreover, since matrix $\mathbf{B}$ sums to zero, the total imbalance between clusters is null on average, which results in a "closed" system of reciprocal exchanges between clusters.

Thus, by using (4.8) and (4.9), the approximation skew-symmetric matrix follows

$$\mathbf{K} = \mathbf{K}^W(\mathbf{U}) + \mathbf{K}^B(\mathbf{U}) = (\mathbf{W}\mathbf{U}' - \mathbf{U}\mathbf{W}') + (\mathbf{B}(\mathbf{1}_{nK} - \mathbf{U})' - (\mathbf{1}_{nK} - \mathbf{U})\mathbf{B}'). \tag{4.10}$$

The model

Finally, the model which jointly fits the symmetric and skew-symmetric components of the asymmetric dissimilarities $\boldsymbol{\Omega}$ is fully specified as

$$\begin{aligned}\boldsymbol{\Omega} = &\mathbf{U}\mathbf{D}^W\mathbf{U}' - \text{diag}\big(\mathbf{U}\mathbf{D}^W\mathbf{U}'\big) + \mathbf{U}\mathbf{D}^B\mathbf{U}' + \big(\mathbf{W}\mathbf{U}' - \mathbf{U}\mathbf{W}'\big) \\ &+ \big[\mathbf{B}(\mathbf{1}_{nK} - \mathbf{U})' - (\mathbf{1}_{nK} - \mathbf{U})\mathbf{B}'\big] + \mathbf{E}\end{aligned} \tag{4.11}$$

subject to

- $\mathbf{U}$ binary membership matrix
- $\mathbf{D}^B$ ultrametric
- $(\mathbf{1}_n'\mathbf{W} = \mathbf{0}_n)$, $(\mathbf{1}_n'\mathbf{B}\mathbf{1}_K = 0)$.

The model is fitted in the least-squares framework, but since the symmetric component is often more prominent than the skew-symmetric one, a weighted loss function is minimized

$$LS(\mathbf{M}, \mathbf{N}, \mathbf{S}, \mathbf{K}) = \frac{1}{\|\mathbf{M}^2\|}\|\mathbf{M} - \mathbf{S}\|^2 + \frac{1}{\|\mathbf{N}^2\|}\|\mathbf{N} - \mathbf{K}\|^2.$$

An appropriate Alternating Least-Squares (ALS) algorithm is presented in Vicari [16].

Parsimonious models

A restricted and simpler model can be derived for the approximation matrix $\mathbf{S}$ by assuming that $\mathbf{D}^W = \pi_1\mathbf{I}_K$ and $\mathbf{D}^B = \pi_2(\mathbf{1}_{KK} - \mathbf{I}_K)$, where $\mathbf{I}_K$ denotes the $(K \times K)$ identity matrix.

Such a partitioning model defines clusters having within dissimilarities all equal to π_1 and separations between clusters all equal to π_2. Since π_1 is an average measure of internal cohesion of clusters and π_2 measures the average separation between clusters, the reasonable constraint $\pi_1 \leq \pi_2$ may be imposed to ensure homogeneous and well-separated clusters and to derive a two-level hierarchy with merging levels π_1 and π_2 [18].

A parsimonious model can be also specified for the skew-symmetric component. It can be assumed that only one value b_j is present in any column j of $\mathbf{B}$, representing the average imbalance from any object in cluster j to any other object not belonging to cluster j, which in matrix form can be written as $\mathbf{B} = \mathbf{U}diag(\mathbf{b})$, where $\mathbf{b}$ is the n-vector containing the K values b_j. Such a restricted model allows for displaying the results in a joint graphical representation of the symmetric and skew-symmetric components.

One-cluster case

In the special case where only one cluster is required, $\mathbf{U}$ is an n-vector of ones, the weight matrix $\mathbf{B}$ is a vector of zeros, $\mathbf{W}$ reduces to a vector $\mathbf{w}$, which results to be the average imbalance from any objects regardless of the destination: this is just the solution for the Brossier model (Sect. 4.2.2) which does not involve the imbalances in the clustering process.

Application of the model by Vicari [16] *to EuroLang data*

The model for simultaneous clustering of symmetric and skew-symmetric data has been fitted to the EuroLang data (Table 4.1).

For any choice of the number of clusters (from 1 to 5), the best solution in 100 runs from different random starting partitions has been retained, and the solution corresponding to $K = 3$ clusters has been considered where the loss values stabilize. In fact, since the loss decreases with the model complexity (i.e. the number of clusters), the value of K where the loss stabilizes indicates that by further increasing the number of clusters, so little more of the data is explained that no much relevant information is added. Thus, for the sake of parsimony, the solution in K clusters is retained as optimal.

The resulting clusters (Dutch, Engl, Germ), (Rom, Ital, French), (Span) identify groups of languages similar for both average percentages and directions of speaking.

Table 4.5a reports the estimated percentages between clusters which are generally all large mainly due to the contribution of the symmetric part $\widehat{\mathbf{D}}^B$ (Table 4.5b) which measures the separations between clusters, i.e. how clusters of languages are different on average. Table 4.5a reveals that 99.71% of people in (NL, UK, DE) do not speak Spanish; conversely, 94.13% of Spaniards do not master Germanic languages (Dutch, Engl, Germ). From $\widehat{\mathbf{D}}^B$ in Table 4.5b, it can be noted that although (Span) is equally dissimilar (96.92%) on average from (NL, UK, DE) and (RO, IT, FR), when the directions of speaking are also taken into account as in Table 4.5a, then such percentages of non-Spanish speakers are much higher for people who are native Germanic (99.71%) than Romance languages (96.30%) speakers.

Moreover, from $\widehat{\mathbf{D}}^W$ in Table 4.5b the second cluster (RO, IT, FR) results to be more heterogeneous than (NL, UK, DE).

In Table 4.5a, the largest asymmetry is observed between (NL, UK, DE) and (RO, IT, FR) with a stronger tendency of the former to master less the Romance languages than the reverse.

Figure 4.7 displays the simultaneous representation of the results and reflects what is observed: in the tree the heights where the clusters merge are more distant between the first two clusters, while Spanish is less different from the Romance languages, as expected.

As for the component within clusters, the lengths of the leaves in the trees are proportional to the average skew-symmetries, and it is clear that the asymmetric component of the data is much less relevant than the symmetric one.

As far as the language interactions within clusters, the *within* weights in $\widehat{\mathbf{W}}$ (Table 4.5c) provide the different lengths of the leaves (red vertical lines). Each red line below zero (due to a positive weight) means that the percentage of people non-speaking languages in country i is larger than the percentage of people who do not master foreign language i within the same cluster: in cluster 1, the percentage

Table 4.5 EuroLang data: three-cluster solution from the joint model by Vicari [16]

(a) Estimated percentages of non-speaking languages

$\left(\widehat{\mathbf{S}}+\widehat{\mathbf{K}}\right)$	Dutch, Engl, Germ	Rom, Ital, French	Span
NL, UK, DE	0	97.35	99.71
RO, IT, FR	90.54	0	96.30
ES	94.13	97.54	0

(b) Estimated separations between clusters and heterogeneities within clusters

	$\widehat{\mathbf{D}}^B$			$\widehat{\mathbf{D}}^W$		
	Dutch, Engl, Germ	Rom, Ital, French	Span	Dutch, Engl, Germ	Rom, Ital, French	Span
NL, UK, DE	0	93.94	96.92	77.00	0	0
RO, IT, FR	93.94	0	96.92	0	96.33	0
ES	96.92	96.92	0	0	0	0

(c) Within and between weights

	$\widehat{\mathbf{B}}$			$\widehat{\mathbf{W}}$		
	Cluster 1	Cluster 2	Cluster 3	Cluster 1	Cluster 2	Cluster 3
DE/Germ	0.43	0	0	−3.67	0	0
ES/Span	0	0	−0.93	0	0	0
FR/French	0	−1.00	0	0	2.83	0
IT/Ital	0	−1.79	0	0	−0.83	0
NL/Dutch	−0.14	0	0	−13.66	0	0
UK/Engl	5.29	0	0	17.33	0	0
RO/Rom	0	−1.86	0	0	−2.00	0

of people in UK who do not know Dutch and German is higher than the reverse (English is the target foreign language in cluster 1).

Conversely, a red line above zero (due to a negative weight) indicates a not widely spoken language; for instance, in cluster 1 almost all Germans and English do not speak Dutch.

To a lesser extent, similar considerations hold within the second cluster for French (target language) and Romanian (nobody speaks Romanian, but a few Romanians speak French).

Note that since matrix $\widehat{\mathbf{W}}$ sums to zero by column, the red branches in Fig. 4.7 are proportional to the within skew-symmetries, but on average they do not influence the lengths of the between dissimilarities.

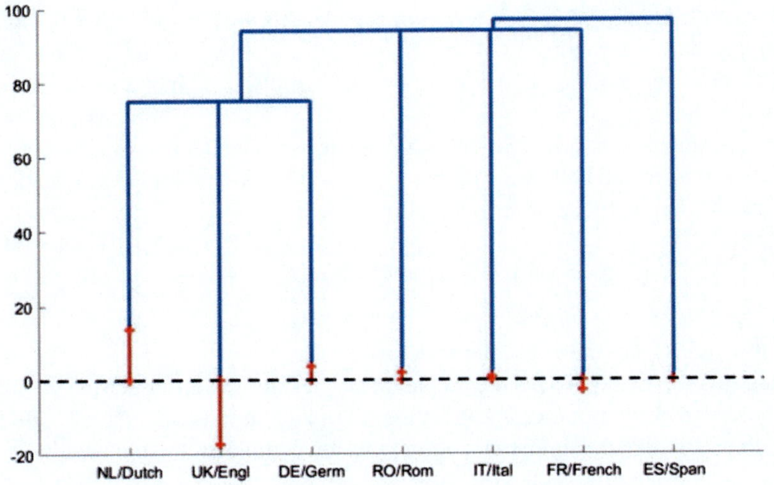

Fig. 4.7 Additive tree from the joint model by Vicari [16] for EuroLang data

4.3.2 Including External Information: CLUSKEXT Model

In many empirical situations, external information on the objects may be available, i.e. a number of variables not explicitly involved in the process of collecting or computing the proximities themselves may be used as additional information to better interpret and explain the nature and directions of the asymmetric relationships between objects.

For example, in psychometric experiments the features of the stimuli provided by the experimenter can be used in the following analysis to better investigate and interpret the latent psychological structures present in the data; similarly, in the analysis of brand switching data, the features of the products or brands under study can be used to investigate on the determinants of the preferences.

In a confirmatory perspective, a matrix of external variables, say **X**, to be used as covariates can be included into the analysis to assess an implicit expectation about certain properties of the clustering configuration.

Starting from model (4.10), Vicari [19] proposed a CLUstering model for SKew-symmetric data including EXTernal information (CLUSKEXT) which incorporates an $(n \times p)$ matrix **X** of external variables on the objects into the approximating clustering structure **K**.

Note that the approximation skew-symmetric matrix (4.10) can be equivalently rewritten as

$$\begin{aligned}\mathbf{K} &= \left[\text{diag}(\mathbf{w})\mathbf{U}\mathbf{U}' - \mathbf{U}\mathbf{U}'\,\text{diag}(\mathbf{w})\right] \\ &\quad + \left[\text{diag}(\mathbf{b})(\mathbf{1}_{nK} - \mathbf{U})\mathbf{U}' - \mathbf{U}(\mathbf{1}_{nK} - \mathbf{U})'\text{diag}(\mathbf{b})\right],\end{aligned} \tag{4.12}$$

where $\mathbf{w}$ and $\mathbf{b}$ are the vectors of size n of the weights within and between clusters, respectively, and, since matrices $\mathbf{W}$ and $\mathbf{B}$ have the same structure of the membership matrix $\mathbf{U}$ (only one non-null entry per row), they simply correspond to the row sums of $\mathbf{W}$ and $\mathbf{B}$ themselves.

In the presence of external variables $\mathbf{X}$, which can be generally supposed column centred, the within and between weights $\mathbf{w}$ and $\mathbf{b}$ are assumed linearly dependent on the $\mathbf{X}$ variables up to an error term as follows

$$\mathbf{w} = \mathbf{X}\boldsymbol{\beta}_W + \boldsymbol{\varepsilon}_W \quad \text{and} \quad \mathbf{b} = \mathbf{X}\boldsymbol{\beta}_B + \boldsymbol{\varepsilon}_B, \tag{4.13}$$

where $\boldsymbol{\beta}_W$ and $\boldsymbol{\beta}_B$ are the coefficient vectors of size p.

Actually, the p external variables act as covariates in accounting for the variability of the observed skew-symmetries and, plugging (4.13) into (4.12), the CLUSKEXT model derives where the unknown coefficients are estimated and the weight vectors $\widehat{\mathbf{w}}$ and $\widehat{\mathbf{b}}$ are computed straightforwardly.

Note that setting $\mathbf{X} = \mathbf{I}_n$ the unconstrained model (4.10) by Vicari [16] follows, which corresponds to the case where the external variables are neither used nor available.

In order to fit the CLUSKEXT model in the least-squares sense, the same algorithm given in Vicari [16] for the skew-symmetric component can be used to estimate the optimal partition $\widehat{\mathbf{U}}$ and only the steps for updating $\widehat{\mathbf{w}}$ and $\widehat{\mathbf{b}}$ need to be modified to get the estimates $\widehat{\boldsymbol{\beta}}_W$ and $\widehat{\boldsymbol{\beta}}_B$ [19].

In Vicari [19], a useful decomposition of the fit is also provided which, given the optimal partition $\widehat{\mathbf{U}}$ into K clusters, allows to measure the different effects (within and between) and whether the model including the external variables is adequate when compared with the unconstrained model without any external information.

A graphical representation of the estimated skew-symmetries within and between clusters is also presented in Vicari [19] to identify the directions of the imbalances between objects, so that clusters which are mainly origins and destinations of the exchanges are identified, while the determinants of the asymmetries may be interpreted through the estimated within and between regression coefficients.

Application of the CLUSKEXT model to EuroLang data

From the Special Eurobarometer-243 survey [1], many information on citizens' attitudes to multilingualism are available in addition to data in Table 4.1. The samples of Europeans were investigated about their opinions on the importance of learning and speaking foreign languages especially for young people (see Table 1.5).

In order to consider the determinants of the asymmetry of the reciprocal use of foreign languages, the following percentages have been considered as external information (Table 4.6):

X_1—people who learned or improved their first foreign language at secondary school;

Table 4.6 EuroLang data: external variables on countries (Special Eurobarometer-243 [1])

	DE	ES	FR	IT	NL	UK	RO
X_1	69	41	65	28	80	56	45
X_2	84	77	76	66	58	55	72
X_3	67	44	51	41	91	38	47

X_2—people who think it is important that young people learn other languages at school or University for improving job opportunities;
X_3—people speaking at least one foreign language.

The languages have been firstly clustered without using the external variables by fitting model (4.10) to the skew-symmetric component **N** (Table 4.3) of the EuroLang data: for any choice of the number of clusters (from 1 to 5), the best solution in 100 runs from different random partitions has been retained. Since the loss function values stabilize at $K = 3$, the partition into three clusters has been chosen which accounts for 87.26% of the total imbalances and provides the following three clusters: (Germ, Dutch, Engl), (Span), (French, Italian, Rom).

Furthermore, the CLUSKEXT model (4.12) constrained with (4.13) has been also applied by including the (7×3) matrix $\mathbf{X} = [X_1, X_2, X_3]$ in Table 4.6 as external information and, by varying the number of clusters from 1 to 5 and retaining the best solution in 100 runs from different random starts, again the partition into three clusters has been selected as the best one to be analysed. Such a solution accounts for 87% of the total imbalances, and, since what is lost in accounting for the total sum of squares by the constrained model instead of the standard (unconstrained) one is so small, in a confirmatory perspective the CLUSKEXT solution can be regarded as fully satisfactory and the model adequate to fit the data.

The three clusters result exactly the same: (1) Germanic languages, (2) Romance languages and (3) Spanish, as shown in Fig. 4.8, where the vertical axis represents both the estimated between ($\widehat{\mathbf{b}}$) and within ($\widehat{\mathbf{w}}$) weights of the languages.

The asymmetry *within* clusters is represented by the dotted branches above/below the zero line. Since imbalances between languages within clusters are approximated by differences between weights $\widehat{\mathbf{w}}$ according to model (4.12), from Fig. 4.8 it is possible to identify pairs of languages having large positive imbalances: they are approximated by the differences between vertical dotted lines above zero (positive within weights) directed to points below zero (negative within weights).

Hence, in the first (blue) cluster English is the "target" language for Dutch and German native speakers, because the difference between the large positive weight of English and the large negative weight of Dutch gives a large positive imbalance: more people in UK do not master Dutch and German than people in NL and DE for English as a foreign language. Similarly, in the third cluster, French is spoken by Italians and Romanians more than the reverse, albeit to a much lesser extent.

In Fig. 4.8, the asymmetry *between* clusters is displayed by the horizontal lines above/below the zero line, so that the difference between a pair of square points

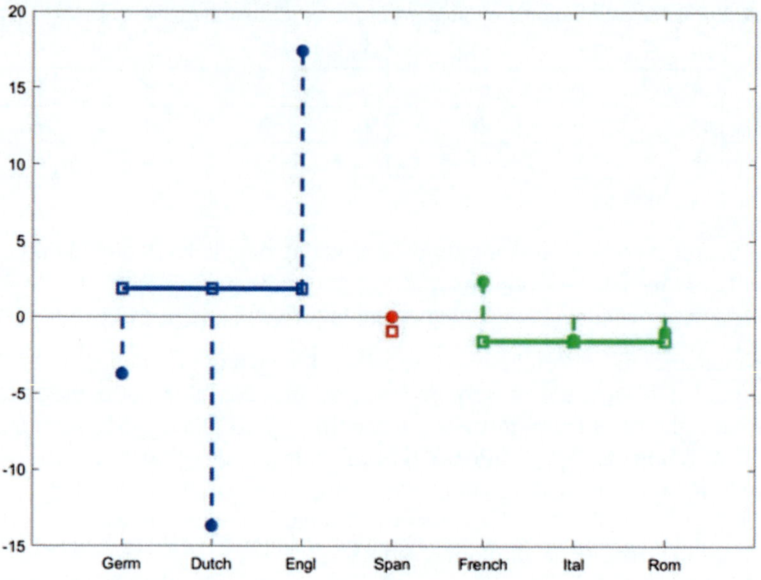

Fig. 4.8 Graph from the CLUSKEXT model for EuroLang data

(identifying a pair of languages) visualizes the imbalance between the corresponding clusters they belong to. Clusters above zero include languages with positive between weights $\widehat{\mathbf{b}}$ and identify countries where foreign languages in clusters below zero are spoken the least. The asymmetry has a clear direction from the group of the Romance languages towards both Spanish and, even more, Germanic languages. In addition to such results, which here are the same of the model without external variables, CLUSKEXT provides also the estimated within and between regression coefficients $\hat{\boldsymbol{\beta}}_W = (0.34, -0.15, -0.73)$ and $\hat{\boldsymbol{\beta}}_B = (0.07, 0.24, 0.17)$ from which the within and between weights are computed, respectively.

The regression coefficients indicate that the major effect on the exchanges *within* clusters is due to the % of people speaking at least one foreign language ($\hat{\beta}_W^{(X_3)} = -0.73$): the larger the percentage, the smaller the asymmetry, while the % of people who improve their language skills at secondary school produces the opposite effect ($\hat{\beta}_W^{(X_1)} = 0.34$). For example, within cluster 1, NL and UK are the countries with the highest and lowest % of people who speak at least one foreign language, respectively.

As for the determinants of the *between* asymmetry, the regression coefficients generally point out that on average the larger the percentages of the external variables, the larger the asymmetry because coefficients are all positive and the main effect is due to the % of people who think that mastering foreign languages improves job opportunities for young people ($\hat{\beta}_B^{(X_2)} = 0.24$).

Finally, to inspect the fit of the optimal solution from the CLUSKEXT model in somewhat more details, the decomposition of the Total Sum of Squares (SS) has been computed to measure the loss due to clustering and regression.

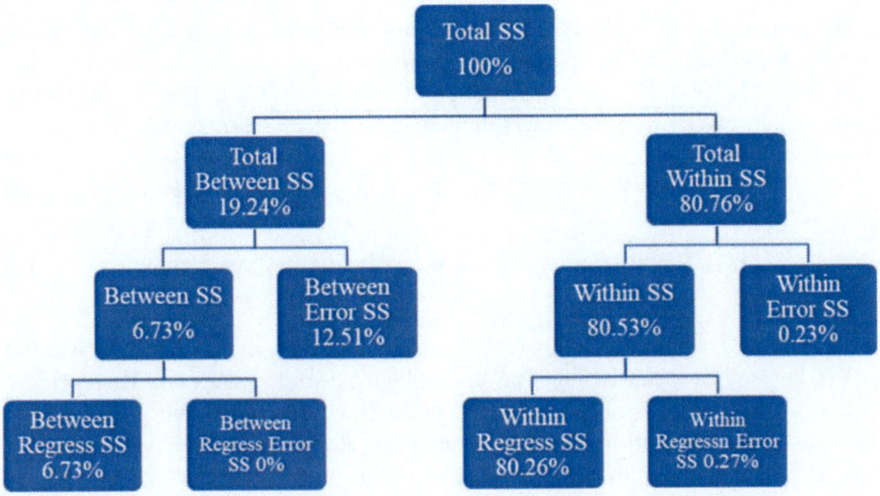

Fig. 4.9 Decomposition of the Sum of Squares for the CLUSKEXT solution for EuroLang data

Specifically, as synthesized in Fig. 4.9 the CLUSKEXT model accounts for about 87% of the Total Sum of Squares, i.e. [Between (6.73%) + Within (80.26%)] Regression SS. Equivalently, the relative loss is about 13%, i.e. [Between (12.51%) + Within (0.23%)] Error SS + [Between (0%) + Within Regression (0.27%)] Error SS. This confirms the goodness of the model where the imbalances between clusters are well explained by the external variables.

For comparison, the corresponding unconstrained model (without external information), for the same optimal partition, accounts for 87.26% of the Total Sum of Squares, i.e. [Between (6.73%) + Within (80.53%)] SS, which is just slightly better than the constrained model and confirms the validity of the results from the covariates.

4.3.3 Modelling Asymmetries Between Clusters

Following the same line of Vicari [16, 19], a non-hierarchical clustering model for asymmetric dissimilarity data $\mathbf{\Omega}$ is presented in Vicari [20], where the symmetric and skew-symmetric components are jointly fitted by two similar approximation matrices depending on two generally different partitions of the objects. Specifically,

1. a "*complete*" partition, where all the n objects are assigned to some cluster, which fits the symmetric component $\mathbf{M}$;
2. an "*incomplete*" partition, where only a subset of objects belong to some cluster and the remaining ones are allowed to remain unassigned, which fits the skew-symmetries $\mathbf{N}$.

The idea is that some objects may have a low degree of asymmetry which can be properly accounted for by the symmetric component.

Matrices $\mathbf{M}$ and $\mathbf{N}$ are approximated by two similar clustering structures depending on two $(n \times K)$ membership matrices $\mathbf{U}$ and $\mathbf{V}$, respectively,

$$\mathbf{M} = \mathbf{S} + \mathbf{E}_S = \mathbf{UC}(\mathbf{1}_{nK} - \mathbf{U})' + (\mathbf{1}_{nK} - \mathbf{U})\mathbf{CU}' + \mathbf{E}_S \tag{4.14}$$

$$\mathbf{N} = \mathbf{K} + \mathbf{E}_K = \mathbf{VD}(\mathbf{1}_{nK} - \mathbf{V})' + (\mathbf{1}_{nK} - \mathbf{V})\mathbf{DV}' + \mathbf{E}_K, \tag{4.15}$$

where $\mathbf{C}$ and $\mathbf{D}$ are diagonal weight matrices of size K subject to constraint $\mathbf{1}_n'(\mathbf{VD})\mathbf{1}_K = 0$ for identifiability reasons, while the error terms $\mathbf{E}_S$ and $\mathbf{E}_K$ represent the part not accounted for by the models. While the ordinary membership matrix $\mathbf{U}$ identifies a complete partition of the n objects into K clusters, the *incomplete* binary membership matrix $\mathbf{V}$ may have a number n_0 ($\leq n$) of zero rows, which correspond to as many objects that remain unassigned to any of the K cluster.

When (4.14) and (4.15) are combined together and an additive constant term b is also added, the model by Vicari [20] is fully specified

$$\begin{aligned}\boldsymbol{\Omega} = \mathbf{M} + \mathbf{N} = \mathbf{S} + \mathbf{K} + b(\mathbf{1}_{nn} - \mathbf{I}_n) + \mathbf{E} = \left[\mathbf{UC}(\mathbf{1}_{nK} - \mathbf{U})' + (\mathbf{1}_{nK} - \mathbf{U})\mathbf{CU}'\right] \\ + \left[\mathbf{VD}(\mathbf{1}_{nK} - \mathbf{V})' + (\mathbf{1}_{nK} - \mathbf{V})\mathbf{DV}'\right] + b(\mathbf{1}_{nn} - \mathbf{I}_n) + (\mathbf{E}_S + \mathbf{E}_K)\end{aligned} \tag{4.16}$$

subject to

- $\mathbf{U}$ binary membership matrix
- $\mathbf{V}$ incomplete binary membership matrix
- $\mathbf{1}_n'(\mathbf{VD})\mathbf{1}_K = 0$.

Actually, model (4.16) is fitted in a least-squares framework, and an efficient ALS algorithm is described in Vicari [20] for estimating the partitions $\mathbf{U}$ and $\mathbf{V}$, the weight matrices $\mathbf{C}$ and $\mathbf{D}$, and the constant term b.

Interestingly, a constrained model may be assumed which links the two partitions fitted to the two independent components $\mathbf{M}$ and $\mathbf{N}$. In such a restricted model, the incomplete partition (identified by $\mathbf{V}$) from the skew-symmetries is assumed nested into the complete partition (identified by $\mathbf{U}$) from the symmetric component, i.e. $\mathbf{V} \subseteq \mathbf{U}$. This means that for any pair of corresponding clusters $\mathcal{V}_j$ and $\mathcal{U}_j$ ($j = 1, \ldots, K$) of the incomplete and complete partitions, respectively, then $\mathcal{V}_j \subseteq \mathcal{U}_j$. Equivalently, if object i belongs to cluster $\mathcal{U}_j$, then only two choices are admissible: object i either belongs to cluster $\mathcal{V}_j$ or remains unassigned.

The model offers an interesting reading of the exchanges between clusters.

1. Any entry c_j of the diagonal matrix $\mathbf{C}$ represents the average amount between cluster $\mathcal{U}_j$ and any other cluster $\mathcal{U}_h$ ($h \neq j$), corrected for the mean of the average exchanges between all clusters different from $\mathcal{U}_j$. This means that large (small) values for c_j denote clusters with large (small) amounts of exchanges on average.

2. Any entry d_j of the diagonal matrix **D** is the weighted average imbalance from all objects i in cluster $\mathcal{V}_j$ (i.e. $i \in \mathcal{V}_j$) directed to all objects h in clusters different from $\mathcal{V}_j$, (i.e. $h \notin \mathcal{V}_j$) corrected for the average imbalance towards cluster $\mathcal{V}_j$. This means that a positive d_j qualifies cluster $\mathcal{V}_j$ as mainly a *destination* of the exchanges, whereas a negative value of d_j qualifies $\mathcal{V}_j$ as an *origin*, so that cluster j includes objects with similar behaviours in terms of directions of the exchanges towards the other clusters.
3. The n_0 unassigned objects generate null estimated imbalances, i.e. they correspond to (almost) symmetric dissimilarities in $\mathbf{\Omega}$.
4. The constant b is the baseline average amount regardless of any clustering, and, since it is added to all entries, it does not affect the approximation of the skew-symmetric component.

Application to EuroLang data

Model (4.16) by Vicari [20] has been fitted to the EuroLang data. The algorithm has been run for a number of clusters K from 1 to 5, and the best solutions in 100 runs starting from different random initial partitions were retained. By graphically analysing the decrease in the loss function that exhibits an elbow at three clusters, such a solution is chosen and only 0.34% of the observed percentages remains not accounted for by the model.

The optimal complete partition $\widehat{\mathbf{U}}$ and the incomplete partition $\widehat{\mathbf{V}}$ (nested into $\widehat{\mathbf{U}}$) differ for two languages:

$\mathcal{U}_1$ = (Span, Ital, Rom), $\mathcal{U}_2$ = (Engl), $\mathcal{U}_3$= (*Germ*, *French*, Dutch).
$\mathcal{V}_1$ = (Span, Ital, Rom), $\mathcal{V}_2$ = (Engl), $\mathcal{V}_3$ = (Dutch).

Table 4.7 reports both the estimated average percentages between clusters, increased by an amount equal to the optimal constant $\left(\hat{\mathbf{S}} + \hat{b}\right)$, and the estimated imbalances between clusters $\widehat{\mathbf{K}}$.

Table 4.7 offers a synthesis and provides an informative insight of the exchanges between clusters of languages.

The baseline average percentage of people non-speaking foreign languages $\hat{b} = 97.17$ is large, and it is added to the estimated average percentages computed from the optimal matrix $\widehat{\mathbf{C}}$ with main diagonal vector $\hat{\mathbf{c}} = (8.03, -17.03, -7.31)$.

Table 4.7 EuroLang data: three-cluster solution from model by Vicari [20]

$\left(\hat{\mathbf{S}} + \hat{b}\right)$	Span, Ital, Rom	Engl	*Germ*, *French*, Dutch	$\widehat{\mathbf{K}}$	Span, Ital, Rom	Engl	Dutch
ES, IT, RO	–	88.17	97.89	ES, IT, RO	–	−14.13	3.97
UK	88.17	–	72.83	UK	14.13	–	18.10
DE, *FR*, NL	97.89	72.83	–	NL	−3.97	– 18.10	–

Moreover, the main diagonal of $\widehat{\mathbf{D}}$, containing the estimated weights for the imbalances between clusters, results $\widehat{\mathbf{d}} = (-2.03, 12.10, -6.00)$. The positive weight for the second cluster qualifies English as "target" or dominant language, in the sense that most people speak English, but native English speakers do not master foreign languages. Conversely, the large negative weight for the third cluster reveals that Dutch people better master foreign languages. A ranking of the clusters also derives from the magnitudes and signs of the resulting weights: (a) from $\widehat{\mathbf{c}}$, clusters are ranked $(\mathcal{U}_1, \mathcal{U}_3, \mathcal{U}_2)$ with respect to the decreasing average percentage of non-spoken languages, regardless of the directions; (b) from $\widehat{\mathbf{d}}$, clusters are ranked $(\mathcal{V}_2, \mathcal{V}_1, \mathcal{V}_3)$, where $\mathcal{V}_2$ is the target language cluster.

Finally, note that German and French remain unassigned in the partition from the skew-symmetries, and, actually, they do not contribute to estimate the imbalances, because they have low degrees of asymmetry.

4.3.4 Dominant Object Approach

Okada and Yokoyama [21] proposed ACLUSKEW (Asymmetric CLUster analysis based on SKEW-symmetry) as a method for clustering asymmetric similarity data $\mathbf{\Omega}$ based on the concept of "dominant object". The underlying rationale is that an object is dominant or salient when its similarities directed to other objects are smaller than the reverse ones. For example, in brand switching, the dominant brand is the most attractive and can lure consumers from the less dominants; or in EuroLang data English is the dominant language because people from all countries speak English, but a few people in UK speak foreign languages.

Actually, in the case of a proximity matrix $\mathbf{\Omega}$ measuring *similarity* relationships, for any pair of objects (i, j) let us consider that ω_{ij} and ω_{ji} reveal the reciprocal position of the two objects: when $\omega_{ij} > \omega_{ji}$, then object j dominates over object i, and when $\omega_{ij} < \omega_{ji}$, then i dominates over j [22]. In the case of proximities that measure *dissimilarities,* the opposite holds.

Since, the skew-symmetric entries of matrix $\mathbf{N}$ are $n_{ij} = (\omega_{ij} - \omega_{ji})/2$ they can be used to rank the influence of the objects and determine the dominant objects.

Given a triplet (i, j, k), suppose that $n_{ik} = n_{ij}$: then, objects j and k equally dominate over object i, while when $n_{ik} > n_{ij}$ the dominance of object k over i is larger than the dominance of object j. The problem is to find a number K of dominant objects and as many clusters which the remaining $(n - K)$ non-dominant objects are assigned to, by maximizing the following goodness-of-fit function

$$F = \sum_{k=1}^{K} \sum_{i \in C_k} \text{sign}(n_{ik})(n_{ik} m_{ik})^2,$$

where sign() is the sign function, C_k denotes cluster k, while n_{ik} and m_{ik} are the entries of the matrices **N** and **M**, respectively.

Note that the imbalance n_{ik} is weighted with the average amount m_{ik}: the larger the average amount, the larger the weight. The effect is that for $n_{ik} = n_{ij}$, when $m_{ik} > m_{ij}$, then the dominance of object k over i and j is due to the greater importance of k over j in terms of average amounts of the exchanges.

Briefly, the algorithm to identify the clusters can be sketched as follows:

0. Set the number of clusters K.
1. Set a subset of K (dominants) objects chosen from the n objects.
2. Assign each of the (n − K) remaining objects to the cluster which maximizes F.

Repeat Steps 1 and 2 for all the $\binom{n}{K}$ combinations of K objects from the n objects and retain the best solution in terms of F value. Note that the number of combinations becomes quickly a large number as the number of objects n increases.

In order to find the correct number of dominant objects (and clusters), the algorithm can be run for increasing values of K, by analysing the best F values which do not necessarily increase with the number of clusters.

In ACLUSKEW, dominant (or central) objects are assumed to be representative of the clusters themselves, so they play the same role as the cluster centroids in the popular k-means algorithm [23] or the medoids (which are actual objects instead of the means of the objects belonging to the same clusters) in the k-medoids/PAM method (see Kaufman and Rousseeuw [24]). Therefore, in such a respect the ACLUSKEW algorithm can be regarded as a k-means-type algorithm designed for one-mode two-way similarities.

Application of ACLUSKEW to EuroLang data

Since the ACLUSKEW method is designed for proximity data which measure similarity relationships, it has been applied to the EuroLang data after subtracting the percentages in Table 4.1 from 100.

By varying the number of clusters from 2 to 5, the two-cluster solution has been retained because it attains the maximum F value. Such a partition (Table 4.8) correctly highlights the role of English as dominant language spoken by people from all other countries. Spanish which forms a cluster of its own has a different behaviour: Spain is the country where fewer people speak a foreign language.

Table 4.8 EuroLang data: two-cluster solution from ACLUSKEW method

Cluster	Dominant language	Dominated languages
1	Engl	Germ, French, Ital, Dutch, Rom
2	Span	

4.4 Clustering Approaches for Two-Mode Two-Way Data*

Clustering methods originally defined for rectangular data matrices (two-mode two-way data discussed in Chap. 2) are not fully described in this chapter, but just referenced for further reading. More in-depth reviews are covered in Chap. 5 of the book on *Data analysis of asymmetric structures* by Saito and Yadohisa [14] and in Bove and Okada [25].

Therefore, some additional approaches are only sketched here which, originally proposed for two-mode two-way data, can be used or adapted to one-mode asymmetric proximities.

Biclustering approach

In two-mode two-way data, the role played by rows and columns is generally different, and in several contexts row and column objects may be jointly clustered to provide an insight of the associations and links between such objects (for instance, consumers and products).

A *biclustering* (alternatively, also referred to as *two-mode clustering, two-mode partitioning, co-clustering*) approach is often useful to simultaneously search for distinct partitions of row and column objects by allowing for different numbers of clusters for the two partitions (for comprehensive reviews, see Madeira and Oliveira [26], Prelić et al. [27] and Wilderjans et al. [28]). As a special case, such methods can be also applied to *square one-mode* two-way proximity data because they handle simultaneously the whole asymmetric matrix.

An especially important aspect of the application of a biclustering approach to a one-mode matrix concerns the handling of the main diagonal, which can be either meaningless or meaningful depending on the type of application. For example, in the case of social network ties, having the main diagonal elements does not make sense (e.g. Are people friends with themselves?), while in confusion data, the main diagonal exemplifies the correct responses to the presented stimulus. Since they are not specifically designed for one-mode data, ordinary biclustering methods do not generally allow to ignore the main diagonal entries. Brusco et al. [29] discussed and compared three different approaches to biclustering methods including the variant to ignore the diagonal entries. A suite of MATLAB programs (available as a web supplement with the original paper) is also provided that implement the three biclustering approaches and a general process for model selection.

Clustering using asymmetric measures

Since the presence of asymmetry in the data may be due to asymmetrical relationships between objects, data may encompass associations of hierarchical nature which should be taken into account.

For example, in text mining problems, Muñoz and Martín-Merino [30] considered that word associations are not symmetric and claimed the need for asymmetric similarity measures.

They defined two asymmetry coefficients which convey the information provided by asymmetry: the first one derived from the fuzzy logic similarity and the second one associated with the Kullback–Leibler divergence.

Following the same logic, in the view of handling data containing intrinsic asymmetric relationships between objects, Olszewski [31, 32] and Olszewski and Šter [33] proposed to extend the popular *k*-means algorithm [23] along a twofold way: the use of either

1. asymmetric dissimilarities instead of the Euclidean distance (for instance, the Kullback–Leibler divergence, or measures involving parameters to tune the degree of asymmetry of the dissimilarity), or
2. symmetric similarities to be "asymmetrized" by employing proper coefficients.

Therefore, the standard *k*-means algorithm is modified by optimizing a weighted loss function where the asymmetric coefficients are attached as weights.

4.5 Applications

In order to better illustrate how the clustering models behave in recovering the underlying structure existing between objects, two real applications are analysed: the Managers data by Krackhardt [34] reported in Table 1.2 and the European Languages data (Eurobarometer-243 survey [1]) presented in Table 1.4.

4.5.1 Managers Data

Graphical representations of the Managers data in Chap. 2 show the existence of directed relationships in asking help and advice between groups of managers at work which reflect their positions within their departments.

Since most methods presented in this chapter are designed for dissimilarity data, the original similarities in Table 1.2 have been naturally transformed by subtracting each count ω_{ij} from the total numbers of managers $(21 - \omega_{ij})$, so that each value measures how many managers responded that manager *i does not go* to manager *j* for help or advice at work.

The transformed proximity matrix has been firstly symmetrized obtaining the symmetric matrix **M** which accounts for 96.96% of the variance of the data, while the skew-symmetric component **N** accounts for the remaining 3.04%, which, albeit small, can contain non-negligible information on the asymmetry of the relationships between managers, as expected.

Symmetry and skew-symmetry have been independently modelled by running the Brossier method (Sect. 4.2.2) which separately approximates the symmetrized proximities **M** by the standard average linkage method and the skew-symmetries **N** by a matrix of rank 2. The resulting additive tree is displayed in Fig. 4.10.

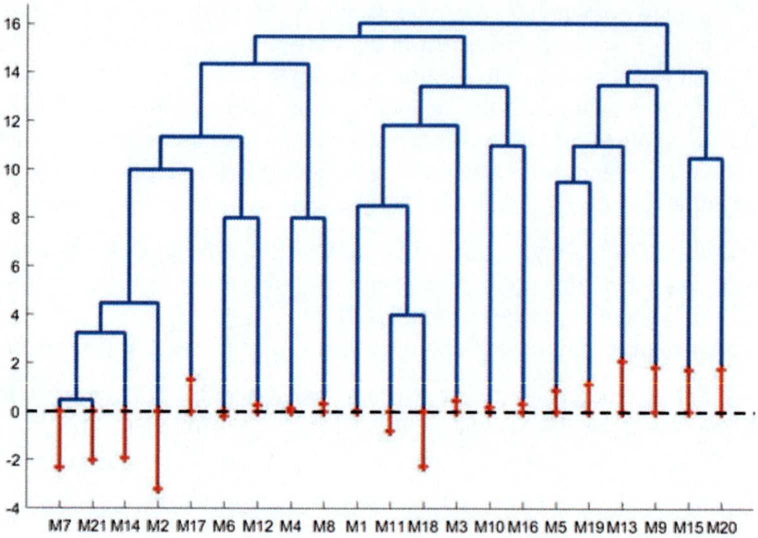

Fig. 4.10 Additive tree from the Brossier method for Managers data

Since the groups are obtained only from the symmetrized dissimilarities, managers in the same groups at low levels have similar average reciprocal requests for help/advice regardless of the directions of such requests. From the lengths of the branches attached to the terminal nodes, it is possible to confirm that the four vice-presidents (M2, M14, M18, M21) and the president (M7) receive more requests than they make, while some supervisors (M13, M9, M15, M20) more often go and ask for advice than give advice.

The model fits well the data because 92.20% of the variance is accounted for.

Because of the independent modelling of the symmetric and skew-symmetric components, it is worth to remind that Fig. 4.10 does not allow to understand "who asks for help from whom", but it is only possible to know the average directions of the requests for each manager. Moreover, the tree is built only from the average data (symmetric part), and the imbalances (skew-symmetric part) do not contribute to the clustering process.

Since such data subsume dominance relations, the ACLUSKEW method by Okada and Yokoyama [21] has been also applied to the original similarities (Table 1.2) by varying the number of clusters from 1 to 8.

The maximum value of the objective function is clearly attained at five clusters (Fig. 4.11) which perfectly reflects the manager hierarchy in terms of role (vice-president/president vs. supervisor) and department (Table 4.9). The dominant managers are the four vice-presidents who are asked for help and advice by the supervisors of the departments they belong to. In addition, the president has a dominant position especially towards supervisor M11 which confirms what was observed in Chap. 2.

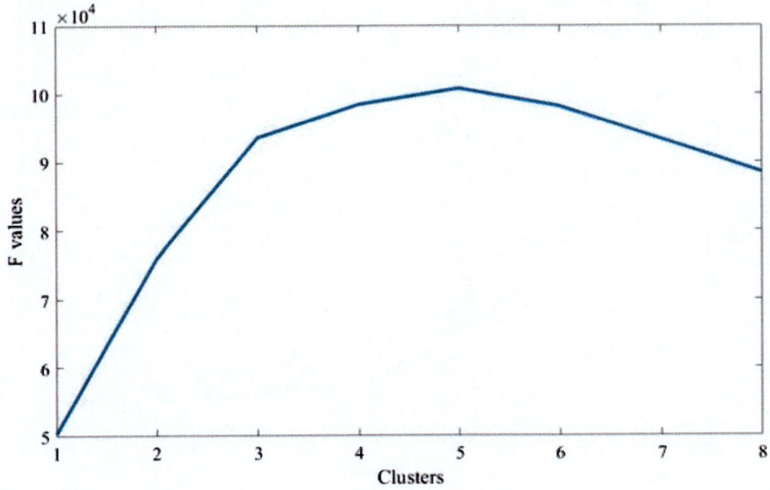

Fig. 4.11 Managers data: objective function versus number of clusters for ACLUSKEW method

Table 4.9 Managers data: five-cluster solution from ACLUSKEW method

Cluster	Dominant manager	Dominated managers
1	M2	M1, M4, M16
2	M7	M11
3	M14	M3, M5, M9, M13, M15, M19, M20
4	M18	M10
5	M21	M6, M8, M12, M17

In order to further shed light on the relationships between managers, the joint model by Vicari [16] presented in Sect. 4.3.1 has been run on the dissimilarity data by varying the number of clusters from 1 to 8 and retaining the best solution in 100 runs of the algorithm for any number of clusters. Even if the "elbow" criterion in the scree plot of the loss function (Fig. 4.12) suggests three clusters, the solution into four clusters, where the loss stabilizes, indicates that no additional relevant information come by increasing the number of clusters. Thus, the four-cluster solution has been analysed in detail which also provides a better insight of the relationships between managers.

The relationships between (groups of) managers are well represented, even if 27% of the variance of the data remains not accounted for.

The four-cluster solution displayed in the tree of Fig. 4.13 identifies Departments 1 (blue cluster) and 2 (purple cluster) with their vice-presidents M21 and M14, respectively, who receive requests for help/advice from supervisors within their departments. Moreover, the first (green) cluster defines the "top managers" because it includes the president M7, who is the main target of the requests, the remaining

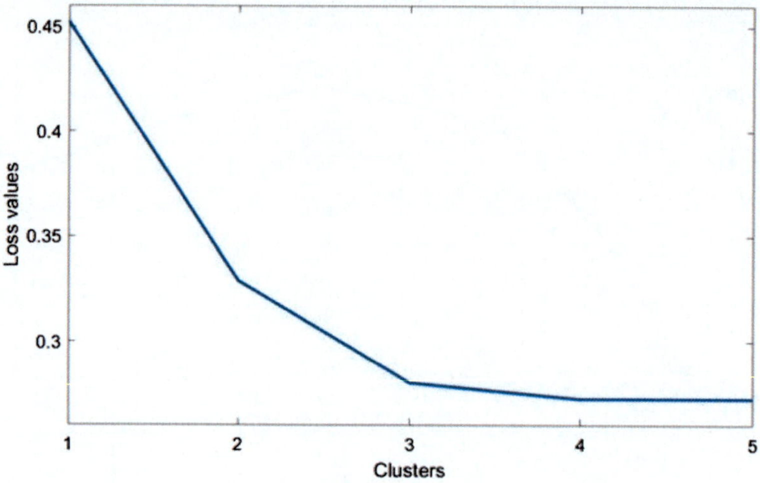

Fig. 4.12 Managers data: loss function versus number of clusters for the joint model by Vicari [16]

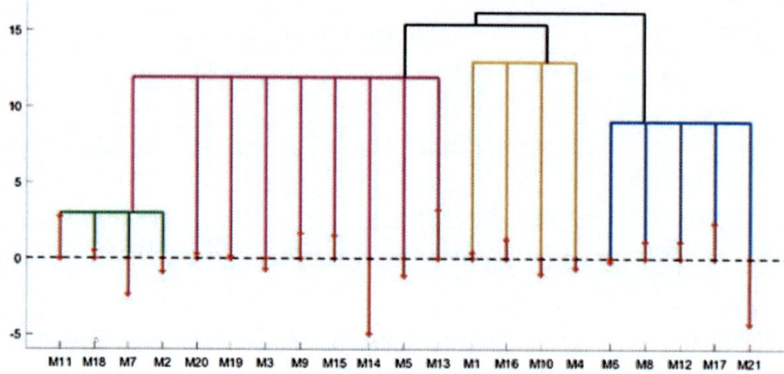

Fig. 4.13 Managers data: additive tree from the joint model by Vicari [16]

two vice-presidents M2 and M18 and supervisor M11, who has a direct relationship with the president. Finally, the third (yellow) cluster corresponds to supervisors in Department 4, in addition to M10.

As far as the relations between groups, note that the different heights of the tree are proportional to the exchanges between clusters of managers where the directions of the requests go from the top to the bottom. Thus, from the tree it is evident that the first (green) cluster results the primary destination of the requests from all departments, as expected, and, to a lesser extent, managers in Department 1 (blue) also receive requests from Departments 2 (purple) and 4 (yellow).

4.5.2 European Languages Data

Another application to real asymmetric data has been considered to better illustrate the performance of some clustering methods.

Data are extracted from the results of the Special Eurobarometer-243 survey [1] on multilingualism in Europe and refer to 16 countries and their languages, including the selected seven countries (*EuroLang* data) already analysed throughout Chap. 4.

Data in Table 1.4 report the percentage ω_{ij} of citizens from country i who responded to speak foreign language j "well enough to have a conversation" and reflect a very asymmetric situation, as expected, with a high asymmetry degree (46.13%).

As a first rough indication of the openness and multilingual ability of the European citizens, Fig. 4.14 reports the average total percentages of both people speaking and language spoken.

English is the most widely foreign language spoken on average, followed by German although to a very much lesser extent. In addition, Swedish, Danish and Dutch people are the most multilingual on average, while Bulgarian and Spanish people are the least ones.

In order to analyse the directions of the linguistic knowledge and detect the possible presence of groups of countries/languages having similar skills, additional information (Table 1.5) coming from several responses to questions about the motivations and the importance of speaking foreign languages has been taken into account for a better understanding of multilingualism.

Fig. 4.14 European Languages data: scatter diagram of average total percentages

As for the EuroLang data, the observed percentages of speaking foreign languages (Table 1.4) have been naturally transformed into dissimilarities by taking the complement to 100 (i.e. $100 - \omega_{ij}$), so that the resulting percentages somehow explain the diversity between native and foreign languages. They have been firstly clustered without using the external variables by fitting the CLUKEXT model (4.10) by Vicari [19].

Since the average percentages present very different magnitudes, the skew-symmetries have been preliminarily weighted by the pairwise symmetries so that any imbalance is related to its order of magnitude ($n^*_{ij} = n_{ij}/m_{ij}$).

For any choice of the number of clusters (from 1 to 6), the best solution in 100 runs from different random starting partitions has been retained, and the scree plot of the loss values against the number of clusters is reported in Fig. 4.15. Since in the case of only one cluster, the loss value (equal to 0.401) is much larger than values for $K \geq 2$, for the clarity of the graphical representation, only the loss values for more than one cluster are shown in Fig. 4.15.

Actually, the elbow of the loss function (not shown in Fig. 4.15) indicates the trivial solution into two clusters where English is separate from the rest of the languages, as expected and evident from the data, but the solution corresponding to $K = 4$ clusters has been considered here where the loss values stabilize (Fig. 4.15).

The four-cluster solution fits well the weighted imbalances because the variance accounted for is 84.43%.

From the graph of the four-cluster solution in Fig. 4.16, it emerges that the distinction is mainly due to the mutual linguistic intelligibility: Romance languages (blue cluster), Germanic languages (red cluster), Slavic languages plus Hungarian (green

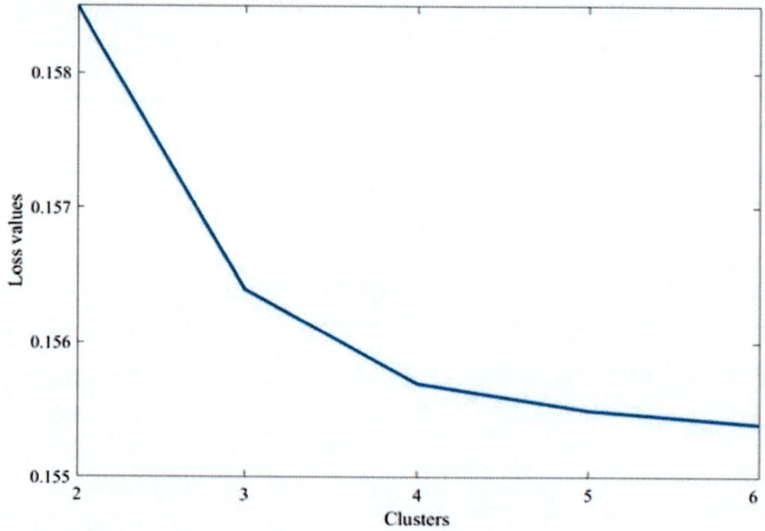

Fig. 4.15 European Languages data: loss function versus number of clusters for CLUSKEXT model

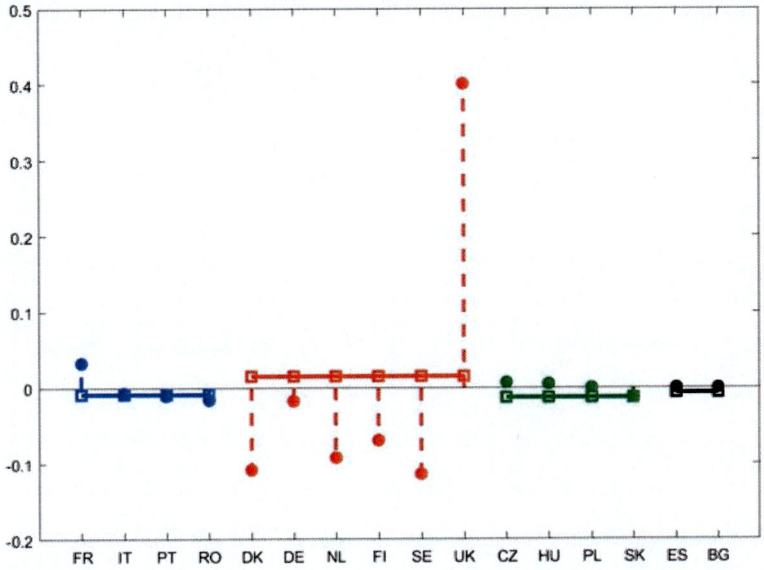

Fig. 4.16 European Languages data: graph of the four-cluster solution from the CLUSKEXT model without external information

cluster) and the last cluster (black cluster) formed by Spanish and Bulgarian. The latter includes the two countries which present the lowest percentages of people speaking foreign languages and whose native languages are the least spoken abroad. By looking at the dashed vertical lines above zero, it follows that within the first two groups the dominant languages are English among the Germanic languages, and French within the group of the Romance languages, even if to a lesser extent. Conversely, people from Sweden, Denmark, The Netherlands and Finland are the best foreign language speakers. Since the heights of the horizontal lines are proportional to the linguistic exchanges between groups with the target languages at the top, it turns out that the second cluster is the most attractive for the presence of English, as expected.

After being centred, the four additional variables of Table 1.5 have been included in the constrained CLUSKEXT model (4.12) and (4.13) as external information and by varying the number of clusters from 1 to 6 and retaining the best solution in 100 runs from different random starts, again the partition into four clusters has been selected as the best one to be analysed. The resulting clusters are exactly the same with non-substantial differences in the estimated weights as evident from Fig. 4.17.

Such a solution accounts for 83.05% of the variance of the weighted imbalances, and, in a confirmatory perspective, it is fully satisfactory because the constrained (parsimonious) model is only 1.38% less explanatory than the unconstrained one. Moreover, it allows to analyse the determinants of the linguistic exchanges through the estimated *within* and

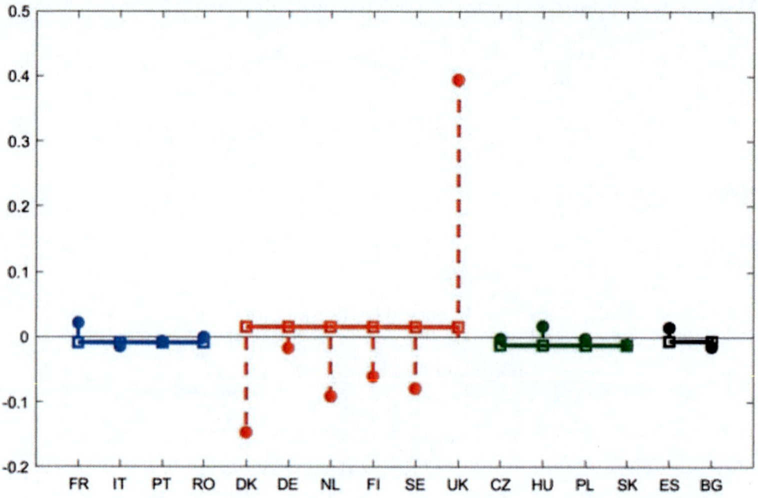

Fig. 4.17 European Languages data: graph of the four-cluster solution from the CLUSKEXT model with external information

between regression coefficients $\hat{\boldsymbol{\beta}}_W$=(−0.0062, −0.0013, 0.0243, −0.0104) and $\hat{\boldsymbol{\beta}}_B$=(−0.0047, −0.0005, 0.0054, 0.0017) from which the within and between weights (proportional to the lengths of the vertical lines in Fig. 4.17) are computed, respectively.

Note that the coefficients have very different magnitudes when compared with the ones estimated for the EuroLang data in Sect. 4.3.2, because in this application the model has been fitted on the weighted skew-symmetries ($n^*_{ij} = n_{ij}/m_{ij}$) which are much smaller; moreover, different information and different countries have been analysed in the two applications.

The main effects of the exchanges *within* clusters are due to (a) how many resident people who speak the language in the country they live as a foreign language ($\hat{\beta}_W^{(X_3)} = 0.0243$): the larger the percentage, the stronger the effect, and to a lesser extent, (b) how many citizens think that learning a new language is useful mainly at work ($\hat{\beta}_W^{(X_4)} = -0.0104$): the larger the percentage the smaller the asymmetry. For example, within the second cluster, DK and UK are the countries with the highest and lowest value of X_4, respectively and correspond to the most and least multilingual countries.

As for the determinants of the *between* asymmetry, the covariates have generally less importance as evident from the graph in Fig. 4.17; the main positive effect is due to ($\hat{\beta}_B^{(X_3)} = 0.0054$), i.e. the percentage of people speaking the language in the country they live as a foreign language (as a proxy of how many foreign people live in the country), while the largest negative effect is due to how many people think that young people should learn other languages for improving job opportunities ($\hat{\beta}_B^{(X_1)} = -0.0047$).

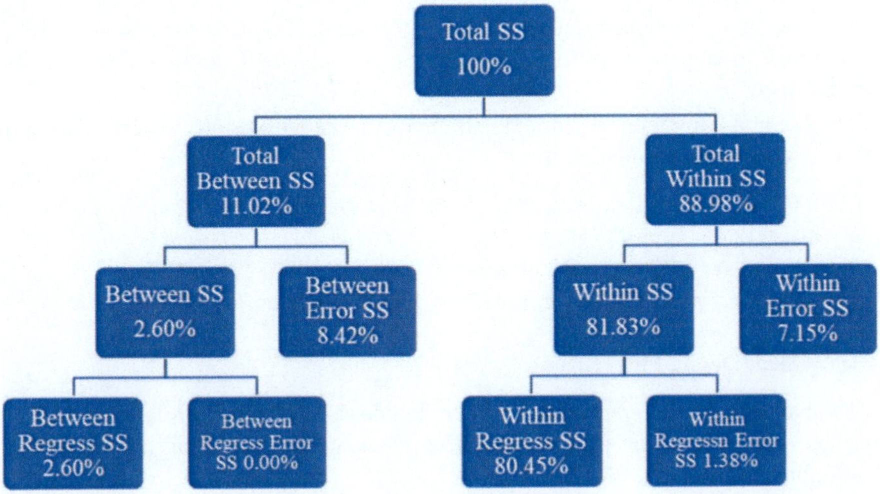

Fig. 4.18 European Languages data: decomposition of the Sum of Squares for the four-cluster CLUSKEXT solution

For example, as far as X_1 concerns, countries in cluster 1 have a slightly larger mean value $\left(\overline{X}_1^{(CL1)}{=}73.5\right)$ than countries in cluster 2 $\left(\overline{X}_1^{(CL2)}{=}71.2\right)$, while the situation is reversed for the average percentage of $X_3\left(\overline{X}_3^{(CL1)}{=}2.5 \text{ and } \overline{X}_3^{(CL2)}{=}4.5\right)$: this results in negative between weights for cluster 1, whereas languages in cluster 2 have positive between weights. Thus, linguistic exchanges are directed from cluster 1 to cluster 2, which means that people from countries in cluster 1 master languages of countries in cluster 2 more than the reverse.

Finally, in order to inspect the fit of the solution from the CLUSKEXT model, Fig. 4.18 reports the decomposition of the Total Sum of Squares (SS) which has been computed to measure the loss due to clustering and regression.

The CLUSKEXT model accounts for 83.05% of the Total Sum of Squares, i.e. [Between (2.60%) + Within (80.45%)] Regression SS. Equivalently, the relative loss is 16.95% which corresponds to the [Between (8.42%) + Within (7.15%)] Error SS + [Between (0%) + Within (1.38%)] Regression Error SS. Note that, given the clustering obtained, when the external information is not used, the relative loss is 15.57%, i.e. [Between (8.42%) + Within (7.15%)] Error SS.

Model (4.16) by Vicari [20] has been also fitted to the European Languages data to provide a different insight of the reciprocal attitudes of the countries to master foreign languages.

The algorithm has been applied to the percentages of non-spoken foreign languages $(100 - \omega_{ij})$ by varying the number of clusters from 2 to 6 and retaining the best solutions in 100 runs from different random initial partitions.

The loss function values stabilize at the three-cluster solution for which 99.73% of the transformed percentages is accounted for, or equivalently, only 0.27% remains not explained.

The optimal complete partition $\widehat{\mathbf{U}}$ and the incomplete partition $\widehat{\mathbf{V}}$ (nested into $\widehat{\mathbf{U}}$) differ only for two languages:

$\mathcal{U}_1 =$ (Czech, Span, Ital, Hung, Pol, Port, Slov, Bulg, Rom),
$\mathcal{U}_2 =$ (Engl),
$\mathcal{U}_3 =$ (Dan, *Germ, French,* Dutch, Finn, Swed);
$\mathcal{V}_1 =$ (Czech, Span, Ital, Hung, Pol, Port, Slov, Bulg, Rom),
$\mathcal{V}_2 =$ (Engl),
$\mathcal{V}_3 =$ (Dan, Dutch, Finn, Swed).

The baseline average percentage of people non-speaking foreign languages $\hat{b} = 99.01$ is very large which confirms that Europeans generally exhibit a low level of multiligualism.

The estimated average percentages are computed from the optimal weights $\hat{\mathbf{c}} = (10.23, -19.80, -10.38)$, while the estimated weights for the imbalances between clusters result $\widehat{\mathbf{d}} = (-0.71, 16.11, -2.42)$. As expected, the large positive weight in $\widehat{\mathbf{d}}$ qualifies English as the "target" or dominant language, because most Europeans master English much more than British people speak foreign languages. Moreover, the negative weight for the third cluster reveals that Danish, Dutch, Finnish and Swedish people are the best in speaking foreign languages.

Finally, note that German and French remain unassigned in the partition from the skew-symmetries and do not contribute to the estimated skew-symmetries.

References

1. European Commission. (2006). *Eurobarometer: Europeans and their languages. Special Eurobarometer 243, Public Opinion Analysis sector of the European Commission, Brussels.*
2. McQuitty, L. L. (1967). Expansion of similarity analysis by reciprocal pairs for discrete and continuous data. *Educational and Psychological Measurement, 27*, 253–255.
3. Hartigan, J. A. (1975). *Clustering algorithms.* Wiley.
4. Landau, S., & Chis Ster, I. (2010). Cluster analysis: Overview. In P. Peterson, E. Baker, & B. McGaw (Eds.), *International Encyclopedia of Education* (3rd ed., pp. 72–83). Elsevier.
5. Everitt, B., Landau, S., Leese, M., & Stahl, D. (2011). *Cluster analysis* (5th ed.). Wiley.
6. Hubert, L. J. (1973). Min and max hierarchical clustering using asymmetric similarity measures. *Psychometrika, 38*, 63–72.
7. Fujiwara, H. (1980). Hitaisho sokudo to toshitsusei keisuu o mochiita kurasuta bunsekiho [Methods for cluster analysis using asymmetric measures and homogeneity coefficient]. *Kodo Keiryogaku [Japanese Journal of Behaviormetrics], 7*(2), 12–21. (in Japanese).
8. Okada, A., & Iwamoto, T. (1995). Hitaisho kurasuta bunnsekihou niyoru daigakushinngaku niokeru todoufukennkann no kanren no bunseki [An asymmetric cluster analysis study on university enrolment flow among Japanese prefectures]. Riron to Houhou [*Sociological Theory and Methods, 10*, 1–13 (in Japanese).
9. Lance, G. N., & Williams, W. T. (1966). A generalized sorting strategy for computer classifications. *Nature, 212*, 218.

10. Lance, G. N., & Williams, W. T. (1967). A general theory of classificatory sorting strategies: 1 Hierarchical systems. *The Computer Journal, 9*, 373–380.
11. Yadohisa, H. (2002). Formulation of asymmetric agglomerative hierarchical clustering and graphical representation of its result. *Bulletin of the Computational Statistics Society of Japan, 15*, 309–316. (in Japanese).
12. Takeuchi, A., Saito, T., & Yadohisa, H. (2007). Asymmetric agglomerative hierarchical clustering algorithms and their evaluations. *Journal of Classification, 24*, 123–143.
13. Okada, A., & Iwamoto, T. (1996). University enrollment flow among the Japanese prefectures: a comparison before and after the joint first stage achievement test by asymmetric cluster analysis. *Behaviormetrika*, 23, 169–185.
14. Saito, T., & Yadohisa, H. (2005). *Data analysis of asymmetric structures. Advanced approaches in computational statistics*. Marcel Dekker.
15. Brossier, G. (1982). Classification hiérarchique à partir de matrices carrées non symétriques. *Statistique et Analyse des Données, 7*, 22–40.
16. Vicari, D. (2014). Classification of asymmetric proximity data. *Journal of Classification, 31*, 386–420.
17. Vichi, M. (2008). Fitting semiparametric clustering models to dissimilarity data. *Advances in Data Analysis and Classification, 2*, 121–161.
18. Vicari, D., & Vichi, M. (2000). Non-hierarchical classification structures. In W. Gaul et al. (Eds.), Data analysis. *Studies in classification data analysis and knowledge organization* (pp. 51–66). Springer.
19. Vicari, D. (2018). CLUSKEXT: CLUstering model for SKew-symmetric data including EXTernal information. *Advances in Data Analysis and Classification, 12*, 43–64.
20. Vicari, D. (2020). Modelling asymmetric exchanges between clusters. In T. Imaizumi, A. Nakayama, & S. Yokoyama (Eds.), Advanced studies in behaviormetrics and data science. *Behaviormetrics: Quantitative approaches to human behavior* (pp. 297–313). Springer Nature.
21. Okada, A., & Yokoyama, S. (2015) Asymmetric CLUster analysis based on SKEW-Symmetry: ACLUSKEW. In I. Morlini, T. Minerva, & M. Vichi (Eds.), Advances in statistical models for data analysis. *Studies in classification, data analysis, and knowledge organization* (pp. 191–199). Springer.
22. Tversky, A. (1977). Features of similarity. *Psychological Review, 84*, 327–352.
23. MacQueen, J. (1967). Some methods for classification and analysis of multivariate observations. In L. M. Le Cam & J. Neyman (Eds.), *Proceedings of the Fifth Berkeley Symposium on Mathematical Statistics and Probability* (Vol. 1, pp. 281–297). University of California Press.
24. Kaufman, L., & Rousseeuw, P. J. (2009). *Finding groups in data: An introduction to cluster analysis*. Wiley.
25. Bove, G., & Okada, A. (2018). Methods for the analysis of asymmetric pairwise relationships. *Advances in Data Analysis and Classification, 12*, 5–31.
26. Madeira, S. C., & Oliveira, A. L. (2004). Biclustering algorithms for biological data analysis: A survey. *IEEE Transactions in Computational Biology and Bioinformatics, 1*, 24–45.
27. Prelić, A., Blueler, S., Zimmermann, P., Wille, A., Bühlmann, P., Gruissem, W., & Zitzler, E. (2006). A systematic comparison and evaluation of biclustering methods for gene expression data. *Bioinformatics, 22*, 1122–1129.
28. Wilderjans, T. F., Depril, D., & Van Mechelen, I. (2013). Additive biclustering: A comparison of one new and two existing ALS algorithms. *Journal of Classification, 30*, 56–74.
29. Brusco, M. J., Doreian, P., & Steinley, D. (2016). Biclustering methods for one-mode asymmetric matrices. *Behavior Research Methods, 48*, 487–502.
30. Muñoz, A., & Martín-Merino, M. (2002). New asymmetric iterative scaling models for the generation of textual word maps. In *Proceedings of the International Conference on Textual Data Statistical Analysis JADT 2002* (pp. 593–603).
31. Olszewski, D. (2011). Asymmetric k-means algorithm. In A. Dovnikar, U. Lotrič, & B. Ster (Eds.), International conference on adaptive and natural computing algorithms (ICANNGA 2011) part II, *Lecture notes in computer science* (Vol. 6594, pp. 1–10). Springer.

32. Olszewski, D. (2012). K-means clustering of asymmetric data. In E. Corchado et al. (Eds.), Hybrid artificial intelligent systems 2012, part I, *Lecture notes in computer science* (Vol. 7208, pp. 243–254). Springer.
33. Olszewski, D., & Šter, B. (2014). Asymmetric clustering using the alpha-beta divergence. *Pattern Recognition, 47*, 2031–2041.
34. Krackhardt, D. (1987). Cognitive social structures. *Social Networks, 9*, 109–134.

Chapter 5
Multiway Models

5.1 Preliminaries

Data covered in the previous chapters can be arrayed in a table which has a set of rows and a set of columns as in Table 2.1. In the table, the three rows correspond to students s_1, s_2, and s_3, and the two columns correspond to ratings of mathematical aptitude: Maths Grade and Maths Self-concept (objective and subjective). The cell (i, j) of the table shows the j-th rating of mathematical aptitude (the rating is objective when $j = 1$ and is subjective when $j = 2$) of student i ($i = 1, 2$, and 3). Another example is Table 2.4 where rows correspond again to students s_1, s_2 and s_3, and columns also correspond to the same set of three students. The cell (i, j) of the table shows the number of peers who considered the i-th student better in Maths aptitude than the j-th student or the dissimilarity between the i-th and j-th student in the mathematical aptitude.

In Table 2.1, rows and columns correspond to different sets: rows correspond to students and columns correspond to ratings on mathematical aptitude. Thus, Table 2.1 represents the values of the three students along two variables, and, in some way, they measure the relationships between a set of students and a set of ratings on mathematical aptitude. Conversely, in Table 2.4 rows and columns correspond to the same set (students s_1, s_2 and s_3): Table 2.4 represents the relationships between students.

Carroll and Arabie [1] introduced a criterion to classify different types of data by using two notions: the mode and the way [2]. The mode indicates a set of objects, subjects, variables, stimuli, occasions, conditions, individuals, points of time, participants and so on. Data represented in Table 2.1 have two modes, one is the set of students and the other is the set of ratings of mathematical aptitude, and are formed by the Cartesian product of such two modes. Data represented in Table 2.4 have only one mode, the set of students, and the mode is replicated. The data are formed by the Cartesian product of the mode (students) by itself. The number of modes (some of them may be repeated) which defines the Cartesian product is the number of ways.

G. Bove et al., *Methods for the Analysis of Asymmetric Proximity Data*, Behaviormetrics: Quantitative Approaches to Human Behavior 7,
https://doi.org/10.1007/978-981-16-3172-6_5

Carroll and Arabie [1] termed the data formed by I modes and J ways as I-mode J-way data. The data represented in Table 2.4 are one-mode two-way data, because the data have one mode and two ways resulted from repeating the mode. The data represented in Table 2.1 are two-mode two-way data, because the data have two modes and have two ways (neither of the two modes is repeated).

In this chapter, we focus our attention on the data having three or more ways which are called *multiway data*, and models dealing with such a kind of data are called *multiway model*. Multiway models of multidimensional scaling to deal with asymmetric multiway data will be described in the present chapter. Unfortunately, there are practically no models and methods of cluster analysis to analyse asymmetric multiway data. We are concerned only with three-way models and methods of multidimensional scaling which, if we may exaggerate, are the only one kind of multiway data and models implemented practically and effectively in the framework of multiway data and models dealing with asymmetric proximity data.

Three-way data are graphically represented by a rectangular solid. There are three types of three-way data: one-mode three-way data, two-mode three-way data and three-mode three-way data. One-mode three-way data have one mode, and the mode is repeated three times which results in three ways. All three ways originate from the same mode. Let us suppose a mode corresponds to a set of n objects, then the relationships of three objects, each from one different way, can be represented by an $(n \times n \times n)$ rectangular solid or cube (Fig. 5.1a). The cell (i, j, k) of the cube represents the relationship of the three objects i, j and k, which can be, for example, a proximity measure of three objects, such as the frequency that three brands are purchased at a same purchasing occasion of a consumer.

Two-mode three-way data have two modes, and one of the modes is repeated. The two ways originate from the same mode, and the third way originates from another not repeated mode. Let us suppose the mode, which is repeated, corresponds to a set of n objects, and the other mode (not repeated) corresponds to a set of K individuals; then, the relationships of two objects and an individual are represented by an $(n \times n \times K)$ rectangular solid or a set of K regular square $(n \times n)$ tables. The cell (i, j, k) of the rectangular solid represents the relationship of objects i and j from the repeated mode and object k from the other mode (not repeated). For example, dissimilarity between two students i and j on the subjective mathematical aptitude at semester k (Fig. 5.1b).

Three-mode three-way data have three different modes, and none of them is repeated. All three ways originate from three different modes respectively. Let us suppose the three modes correspond to a set of n_1 individuals, a set of n_2 variables and a set of n_3 occasions, respectively. The relationships of the n_1 individuals, n_2 variables and n_3 occasions are represented by an $(n_1 \times n_2 \times n_3)$ rectangular solid (Fig. 5.1c). The cell (i, j, k) of the rectangular solid represents the relationship of objects i, j and k, each from a different set of objects, such as, for example, the rating of Maths Grade j of student i at semester k.

As in the previous chapters, the asymmetry in one-mode two-way data means that the relationship between objects i and j is not equal to the relationship between objects j and i or $\omega_{ij} \neq \omega_{ji} (i \neq j)$. For example, the number of people who moved

Fig. 5.1 Three-way data

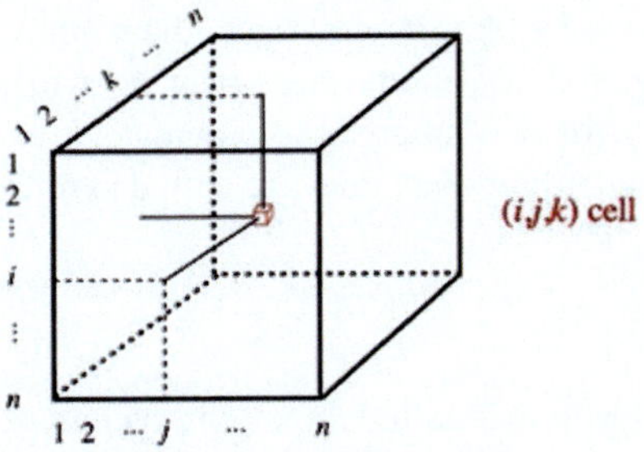

(a) One-mode three-way data

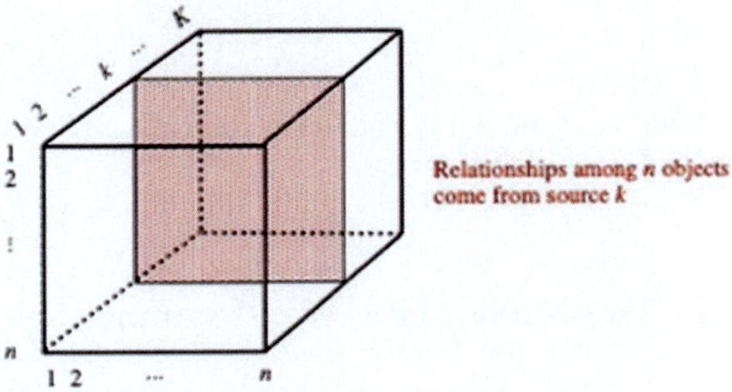

(b) Two-mode three-way data

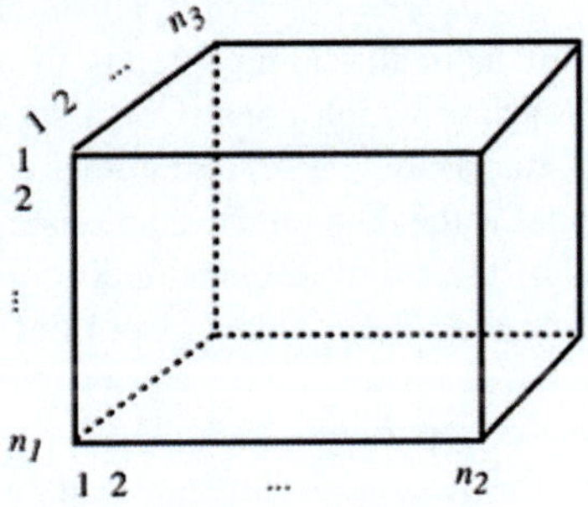

(c) Three-mode three-way data

out from city i towards city j is not necessarily equal to those who moved out from city j towards city i. There is no point in thinking of asymmetry for two-mode two-way data, because it is difficult to define the asymmetry of the relationship between two objects belonging to different sets (different modes). One-mode two-way data can be regarded as two-mode two-way data, when we assume that the two ways originate from two different modes or two different sets of objects (see Sects. 2.3.2.1 and 5.2.2.4). Similarly to two-mode two-way data, it is also meaningless to think of asymmetry for three-mode three-way data.

In the case of three-way data, there are two kinds of asymmetry. One is the asymmetry for one-mode three-way data (cf. Sect. 5.2.2.4), and the other is the asymmetry for two-mode three-way data.

One-mode three-way data are called symmetric [3, 4] when

$$\omega_{ijk} = \omega_{ikj} = \omega_{jik} = \omega_{jki} = \omega_{kij} = \omega_{kji} \tag{5.1}$$

is satisfied, which is called *three-way symmetry* [5]. One-mode three-way data which do not satisfy Eq. (5.1) are called asymmetric one-mode three-way data. Two-mode three-way data, similarly to one-mode two-way data (see Sect. 1.2), are symmetric when

$$\omega_{ijk} = \omega_{jik}, \quad (i, j = 1, 2, \ldots, n; k = 1, 2, \ldots, K). \tag{5.2}$$

Two-mode three-way data which do not satisfy Eq. (5.2) are called asymmetric two-mode three-way data.

5.2 Two-Mode Three-Way Asymmetric Methods

When Tucker and Messick [6] developed their pioneering multidimensional scaling model (point of view model) to analyse two-mode three-way data, they called their model *individual differences model*. This is because two-mode three-way data, which their multidimensional scaling intends to analyse, are object (n) $\times$ object (n) $\times$ individual (K) data which consist of a set of K matrices of proximities among n objects, each matrix is ($n \times n$) and comes from an individual. The shaded square in Fig. 5.1b contains the data on the relationships among n objects coming from source k. The focus of their multidimensional scaling is to disclose the differences of the relationships among the n objects from one individual to another, or the individual differences.

If we analyse separately each of the ($n \times n$) matrix (one-mode two-way data) which forms the two-mode three-way data by using multidimensional scaling or cluster analysis for one-mode two-way data (see Chaps. 2, 3 and 4), it requires to iterate K separate analyses. The K results obtained by the separate analyses can show the individual differences among the K individuals anyway, but they cannot disclose the individual differences in a structured form.

The seminal work on multidimensional scaling for analysing two-mode three-way data was introduced by Carroll and Chang [7] and called INDSCAL (**IN**dividual **D**ifferences **SCAL**ing). Tucker and Messick [6] and Carroll and Chang [7] enable us to analyse simultaneously all the K proximity matrices ($n \times n$) as one set or a group of data, or, equivalently, to analyse two-mode three-way data all together, which made it easy to disclose individual differences.

5.2.1 INDSCAL Model

While the INDSCAL model [7, 8] deals with two-mode three-way symmetric proximities where the $(n \times n)$ proximity matrix is symmetric for all the K individuals, the models introduced thereafter to analyse two-mode three-way asymmetric data inherited the idea of INDSCAL in order to represent the differences among individuals. The INDSCAL model will be briefly explained in this subsection, because it pioneered in giving idiosyncratic set of weights to dimensions in order to represent the differences of the relationships among the objects from one individual to another.

The INDSCAL model consists of an $(n \times r)$ matrix $\mathbf{X}$ where element x_{is} represents the coordinate of object i along dimension s, and of a $(K \times r)$ matrix $\mathbf{W}$, whose element w_{ks} (≥ 0) represents the weight along dimension s given by individual k. Matrix $\mathbf{X}$ represents the configuration of objects which is the basic pattern common to all K individuals, and it is called *common object configuration*. Matrix $\mathbf{W}$ represents the configuration called *weight configuration* of the individuals. The object configuration for individual k, say $\mathbf{X}_k$, is defined by multiplying dimension s of the common object configuration with $\sqrt{w_{ks}}$

$$\mathbf{X}_k = \mathbf{X}\mathbf{W}_k^{1/2}, \tag{5.3}$$

where $\mathbf{W}_k^{1/2}$ is an $r \times r$ diagonal matrix having $\sqrt{w_{ks}}$ as its s-th diagonal element. This means that dimension s of the common object configuration is expanded or shrunk by $\sqrt{w_{ks}}$ to obtain dimension s of $\mathbf{X}_k$, which is the object configuration for individual k. Then, the Euclidean distance between two points representing object i and object j in the object configuration for individual k is given by

$$d_{ijk} = \sqrt{\sum_{s=1}^{r} w_{ks}\left(x_{is} - x_{js}\right)^2}. \tag{5.4}$$

Figure 5.2 shows the two-dimensional common object configuration $\mathbf{X}$ and the weight configuration $\mathbf{W}$ when the number of objects n is five, and the number of individuals K is four. Specifically, Fig. 5.2a shows the common object configuration $\mathbf{X}$ of the five objects, while Fig. 5.2b displays the weight configuration $\mathbf{W}$ of the four individuals. Figure 5.2c depicts the object configuration for individual k_1 whose weights are $w_{k_1 1} = w_{k_1 2}$, and therefore the ratio between the width-to-height of the configuration is the same as that of the common object configuration shown in Fig. 5.2a. Figure 5.2d displays the object configuration for individual k_2 with weights $w_{k_2 1} < w_{k_2 2}$, and thus the ratio between the width-to-height of the configuration is smaller than that in Fig. 5.2a. Conversely, Fig. 5.2e shows the object configuration for individual k_3 with weights $w_{k_3 1} > w_{k_3 2}$, which gives a width-to-height ratio of the configuration larger than that in Fig. 5.2a. Finally, Fig. 5.2f is the object configuration for individual k_4 with equal weights $w_{k_4 1} = w_{k_4 2}$, but being smaller than weights of

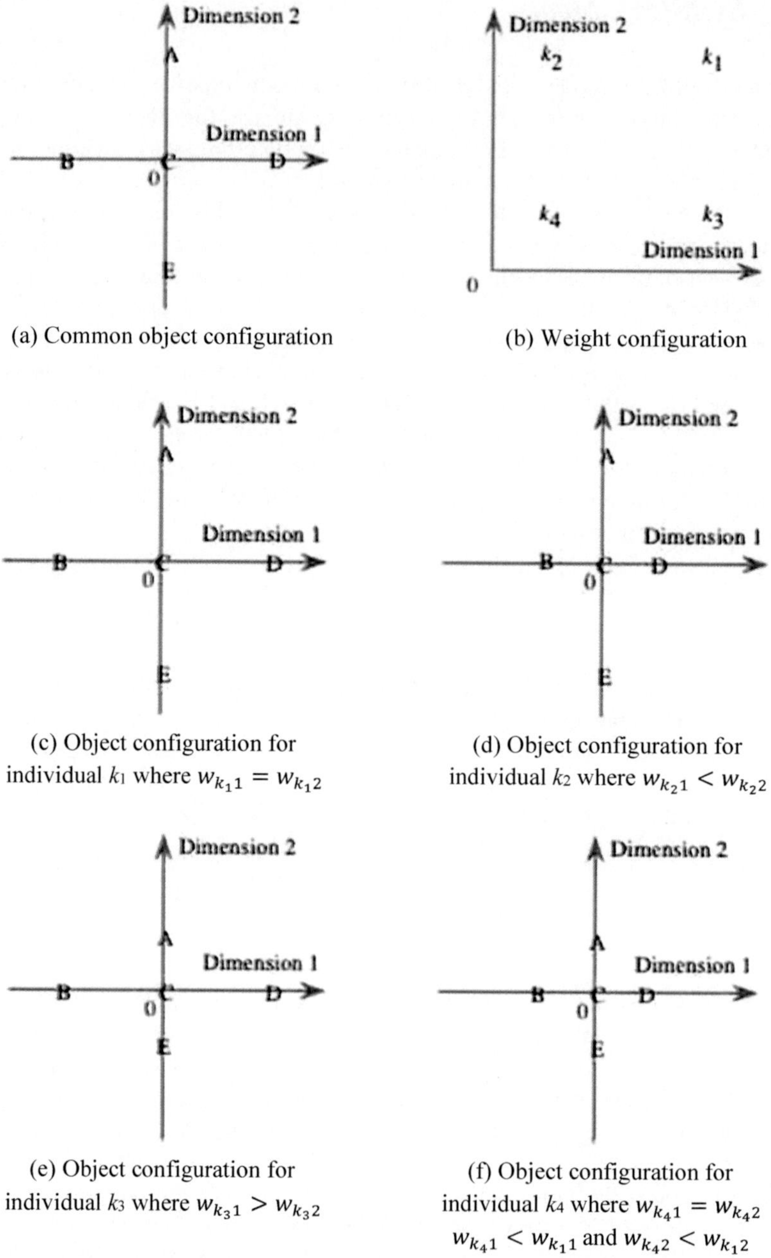

(a) Common object configuration

(b) Weight configuration

(c) Object configuration for individual k_1 where $w_{k_1 1} = w_{k_1 2}$

(d) Object configuration for individual k_2 where $w_{k_2 1} < w_{k_2 2}$

(e) Object configuration for individual k_3 where $w_{k_3 1} > w_{k_3 2}$

(f) Object configuration for individual k_4 where $w_{k_4 1} = w_{k_4 2}$ $w_{k_4 1} < w_{k_1 1}$ and $w_{k_4 2} < w_{k_1 2}$

Fig. 5.2 Two-dimensional INDSCAL model: common object configuration, weight configuration and object configurations for each of four individuals those of individual k_1 ($w_{k_4 1} < w_{k_1 1}$ and $w_{k_4 2} < w_{k_1 2}$), the configuration is uniformly shrunk along dimensions 1 and 2 and geometrically similar to both the common object configuration and to that of individual k_1 shown in Fig. 5.2a and c, respectively. The idea of representing the individual differences of the relationships among objects pioneered by INDSCAL is inherited by several multidimensional scaling models for analysing two-mode three-way asymmetric data.

There are several methods to analyse two-mode three-way asymmetric data which can be classified into two categories based on how they represent the asymmetric relationships among objects. The first one represents *directly* the asymmetric relationships among the objects themselves. The second one represents indirectly the asymmetric relationships among the objects by asymmetric relationships among dimensions where each object has its own set of weights to specify the asymmetric relationships among the objects themselves. In the next two subsections, the two categories of methods for analysing two-mode three-way asymmetric data are described.

For the sake of generality, the broader term *source* instead of *individual* will be used henceforth, with the general meaning of individual, subject, stimulus, occasion, experimental condition, time point, geographical area, situation, scenario and so on. But the term *individual differences* will be used to mean the differences of the relationships among objects from one source to another, because the term individual differences have been used widely for more than five decades.

5.2.2 Representing Asymmetry by Relationships Among Objects

The first category of multidimensional scaling methods is described in this subsection which represent directly the asymmetric relationships among objects by means of the relationships among the objects themselves. The second category of multidimensional scaling method is described in Sect. 5.2.3.

Two major methods belong to the first category, one is the extension of the slide vector model (see Sect. 2.3.2.3; Zielman and Heiser [9–11]), and the other is the extension of the radius-distance model (see Sect. 3.3.1; Okada and Imaizumi [12]). Both of them started from the simpler models proposed for one-mode two-way data.

The two-mode three-way extensions of the one-mode two-way radius-distance model [12] were introduced by Zielman [13] and by Okada and Imaizumi [14] as well. They are illustrated below in Sect. 5.2.2.1 "Radius-distance model (1)" and in 5.2.2.2 "Radius-distance model (2)", respectively.

The two-mode three-way extension of the slide vector model introduced by Zielman and Heiser [9] is presented in Sect. 5.2.2.3.

5.2.2.1 Radius-Distance Model (1)

The radius-distance model, where object i is represented by a point $(x_{i1}, x_{i2}, \ldots, x_{ir})$ and a circle with a radius r_i (in two-dimensional space), a sphere (in three-dimensional space) or a hypersphere (in four- or more-dimensional space) with radius r_i and centred at the point itself in a multidimensional space, is represented by (see Sect. 3.3.1).

$$\omega_{ij} = d_{ij} - r_i + r_j + \varepsilon_{ij}, \tag{5.5}$$

where $d_{ij} = \sqrt{\sum_{s=1}^{r} \left(x_{is} - x_{js}\right)^2}$ is the Euclidean distance between two points representing objects i and j in the r-dimensional configuration of objects, and r_i and r_j are the radii of the circles representing objects i and j, respectively.

The extension of the radius-distance model, originally proposed to analyse one-mode two-way asymmetric data, to two-mode three-way asymmetric data was first introduced by Zielman [13]. Source k has its idiosyncratic object configuration $\mathbf{X}_k$ which is given by multiplying $\mathbf{X}$ with an appropriate transformation matrix $\mathbf{H}_k$

$$\mathbf{X}_k = \mathbf{X}\mathbf{H}_k, \tag{5.6}$$

where $\mathbf{X}$ is the configuration of objects which is common to all sources. The two-mode three-way extension of Zielman [13] is represented by

$$\omega_{ijk} = d_{ijk} + \sum_{s=1}^{r} w_{ks}^{1/2}\left(r_{is} - r_{js}\right) + \varepsilon_{ijk}, \tag{5.7}$$

where d_{ijk} is the Euclidean distance between points representing objects i and j in the object configuration for source k, r_{is} represents the asymmetry of object i along dimension s, and $w_{ks}^{1/2}$ ($w_{ks}^{1/2}$ can be negative) is the weight along dimension s which represents the salience of the asymmetry along dimension s for source k. Positive and negative values for weights $w_{ks}^{1/2}$ denote opposite directions of the asymmetry between two objects. In the object configuration for source k, object i is represented by a point and r ellipses which have the axes of the length $2w_{ks}^{1/2} r_{is}$ along dimension s ($s = 1, 2, \ldots, r$). The model is similar to that of Okada [15] where an object is represented by a point and an ellipse (ellipsoid, hyper ellipsoid). Matrix $\mathbf{H}_k$ can be diagonal with the same effect as $\mathbf{W}_k^{1/2}$ in the INDSCAL model.

Let us suppose $\sqrt{w_{k1}} = \sqrt{w_{k2}} = 1$, then one possible way to geometrically illustrate the model of Zielman [13] is to represent each object as a point and multiple circles centred at that point. The two-dimensional model represents object i as one point and two circles: a circle for dimension 1 with radius r_{i1}, and a circle for dimension 2 with radius r_{i2}. The r-dimensional model represents each object as a point and r circles centred at that point, where the s-th circle

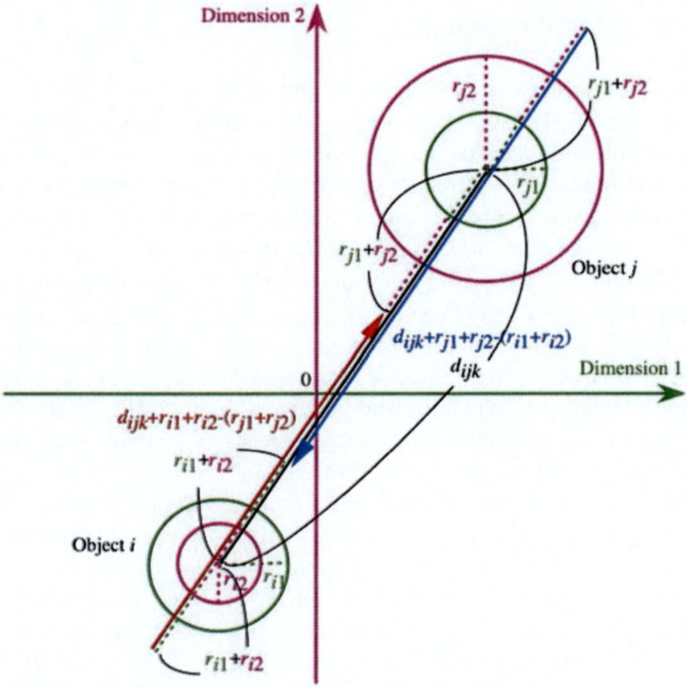

Fig. 5.3 Predicted values of $\omega_{ijk} = d_{ijk}+(r_{i1}+r_{i2})-(r_{j1}+r_{j2})$ and $\omega_{jik} = d_{ijk}+(r_{j1}+r_{j2})-(r_{i1}+r_{i2})$ when $r = 2$ and $\sqrt{w_{k1}} = \sqrt{w_{k2}} = 1$

of object i has radius r_{is} $(s = 1, 2, \ldots, r)$. Figure 5.3 shows objects i and j of a two-dimensional model. The length of the red arrow represents the predicted value $\left(d_{ijk} + (r_{i1} + r_{i2}) - (r_{j1} + r_{j2})\right)$ of ω_{ijk}, and the length of the blue arrow represents the predicted value $\left(d_{jik} + (r_{j1} + r_{j2}) - (r_{i1} + r_{i2})\right)$ of ω_{jik}, when both weights along the two dimensions are assumed to be equal to 1 ($\sqrt{w_{k1}} = \sqrt{w_{k2}} = 1$).

The model represents (a) the individual differences of the symmetric relationships among objects by the transforming matrix $\mathbf{H}_k$, i.e. the first term of the right-hand side of Eq. (5.7), and (b) the individual differences of the asymmetric relationships among objects by means of the weights $w_{ks} (s = 1, 2, \ldots, r)$, i.e. the second term of the right-hand side of Eq. (5.7). The individual differences of the symmetric and asymmetric relationships are independently represented in the configuration.

The algorithm to derive the object configuration, the radii, and the weights is described in detail by Zielman [13] and minimizes the loss function

$$F = \sum_{i=1}^{n} \sum_{j=1}^{n} \sum_{k=1}^{K} \varphi_{ijk} \left\{ \left(\omega_{ijk} - d_{ijk}\right) - \sum_{s=1}^{r} w_{ks} \left(r_{is} - r_{js}\right) \right\}^2, \tag{5.8}$$

where φ_{ijk} is the weight to code possible missing value in the data.

5.2.2.2 Radius-Distance Model (2)

Another two-mode three-way extension of the radius-distance model was introduced by Okada [14]. The model consists of the common object configuration, the asymmetry weight configuration and the symmetry weights. In the common object configuration (common to all sources), each object is represented by a point and a circle [two-dimensional model] (sphere [three-dimensional model], hypersphere [four- or more-dimensional model]) centred at that point. The model is represented by Eqs. (5.12) and (5.13), which was extended from the one-mode two-way model represented by two-way model in Eq. (5.5).

The asymmetry weight configuration represents the weight given to each dimension of the common object configuration to transform a circle (sphere, hypersphere) into an ellipse (ellipsoid, hyperellipsoid) for each source. The symmetry weight is the salience of the symmetric relationships among objects for each source. Let $\mathbf{X} = [x_{is}]$ be the $(n \times r)$ matrix of the r-dimensional configuration of n points each representing an object in the common object configuration, where x_{is} is the coordinate of object i along dimension s $(s = 1, 2, \ldots, r)$. Object i is represented by a point $(x_{i1}, x_{i2}, \ldots, x_{ir})$ and a circle (sphere, hypersphere) with radius r_i in the common object configuration. Matrix $\mathbf{U} = [u_{ks}]$ represents the asymmetry weight configuration, which is the $K \times r$ matrix of the r-dimensional configuration of the K sources, where u_{ks} (≥ 0) is the weight of source k along dimension s which shows the asymmetry of source k along dimension s. The symmetry weight w_k (≥ 0) measures the salience or importance of the symmetric relationships among objects for source k. The model essentially reminds us of the INDSCAL model [7].

Each source k has its own object configuration $\mathbf{X}_k$ which is obtained by multiplying $\mathbf{X}$ with the symmetry weight w_k for source k

$$\mathbf{X}_k = w_k \mathbf{X} \tag{5.9}$$

which uniformly expands or shrinks the common object configuration $\mathbf{X}$ along all $(s = 1, 2, \ldots, r)$ dimensions. In the object configuration for source k, object i is represented by a point $(w_k x_{i1}, w_k x_{i2}, \ldots, w_k x_{ir})$ and an ellipse (ellipsoid, hyperellipsoid) centred at that point; such an ellipse is tailored to source k by u_{ks} and has the axis of length $2u_{ks}r_{is}$ $(s = 1., 2, \ldots, r)$ along dimension s as illustrated in Fig. 5.5. The Euclidean distance between two points representing objects i and j in the object configuration for source k is

$$d_{ijk} = \sqrt{\sum_{s=1}^{r} \left[w_k \left(x_{is} - x_{js} \right) \right]^2} = w_k d_{ij}, \tag{5.10}$$

where

$$d_{ij} = \sqrt{\sum_{s=1}^{r} \left(x_{is} - x_{js}\right)^2}, \tag{5.11}$$

which is the distance between the two points representing objects i and j in the common object configuration.

Figure 5.4 shows the model in a two-dimensional space. Figure 5.4a represents the two-dimensional common object configuration of six objects, each represented by a point and a circle centred at that point (circles of objects other than object i are omitted for clarity). Object i is represented as a point (x_{i1}, x_{i2}) and a circle having radius r_i. Figure 5.4b shows the asymmetry weight configuration representing five sources; specifically, source k is represented by a point (u_{k1}, u_{k2}).

In this model, both symmetric and asymmetric relationships among objects, which are common to all sources, are represented in the common object configuration. The symmetric relationships among objects are described by the interpoint distances d_{ij} $(i, j = 1, 2, \ldots, n : i \neq j)$ in the common object configuration $\mathbf{X}$, while the asymmetric relationships among objects are described by the radii of the circles (spheres, hyperspheres) r_i $(i = 1, 2, \ldots, n)$ in the common object configuration.

The present model accounts for the individual differences of the symmetric relationships among objects by the symmetry weights w_k $(k = 1, 2, \ldots, K)$, and those of the asymmetric relationships among objects by the asymmetry weights u_{ks} $(k = 1, 2, \ldots, K; s = 1, 2, \ldots, r)$. As in the other two-mode three-way extension of the radius-distance model [13], here the individual differences of the symmetric and of the asymmetric relationships are independently represented.

Let ω_{ijk} be modelled as

$$\omega_{ijk} = m_{ijk} + \varepsilon_{ijk}, \tag{5.12}$$

where m_{ijk} is monotonically decreasingly (increasingly) related to ω_{ijk} for source k when ω_{ijk} is a similarity (dissimilarity) and defined by

$$m_{ijk} = d_{ijk} - \nu_{ijk} r_i + \nu_{jik} r_j, \tag{5.13}$$

where ν_{jik} is equal to ν_{ijk} and defined as

$$\nu_{ijk} = \frac{d_{ijk}}{\sqrt{\sum_{s=1}^{r} \left(\frac{x_{is} - x_{js}}{u_{ks}}\right)^2}}. \tag{5.14}$$

When the subscript k and u_{ks} in (5.14) are eliminated, then $\nu_{ijk} = \nu_{jik} = 1$, so that (5.13) practically reduces to (5.5) of the distance-radius model for one-mode two-way data.

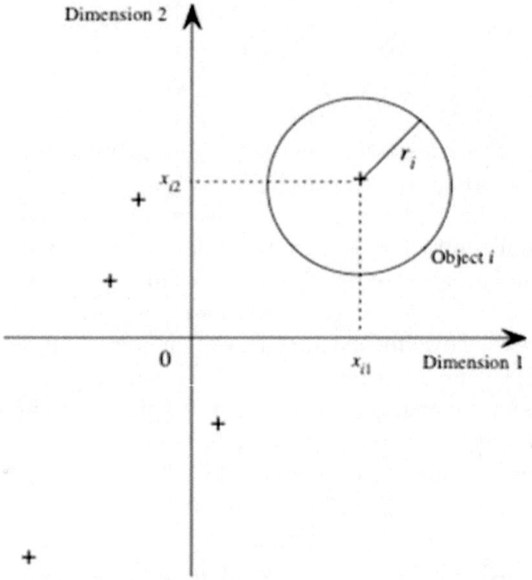

(a) Common object configuration: Six objects are represented in the configuration. Object i is represented as a point (x_{i1}, x_{i2}) and a circle of radius r_i. Circles of other objects are omitted for clarity.

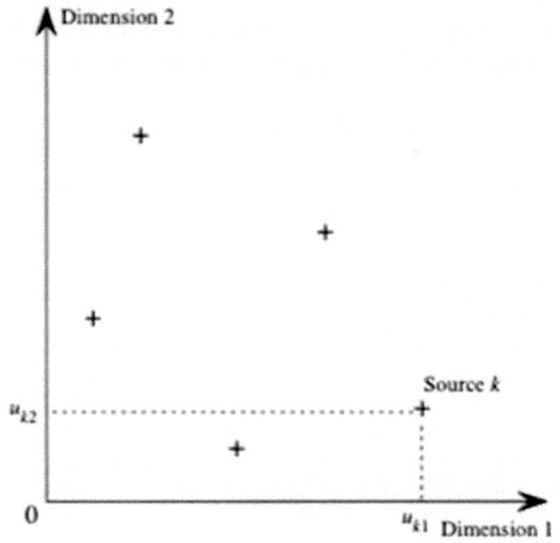

(b) Asymmetry weight configuration: Five sources are represented in the configuration. Source k is represented as a point (u_{k1}, u_{k2}).

Fig. 5.4 Common object configuration and asymmetry weight configuration in a two-dimensional space

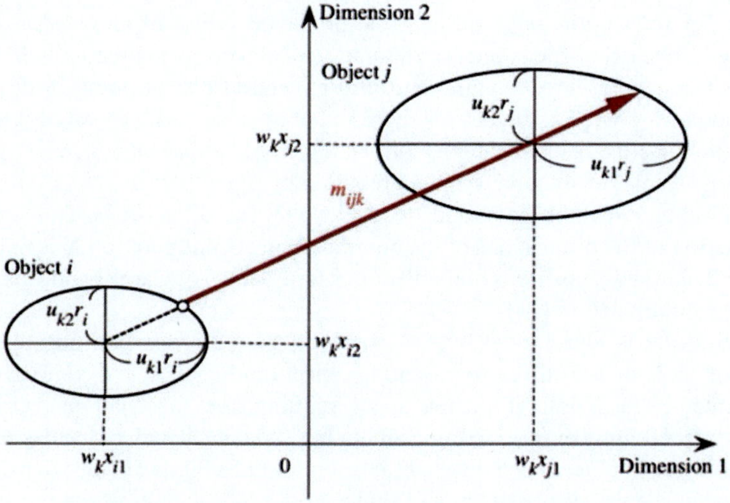

(a) Predicted value of the dissimilarity ω_{ijk} from object i to object j for source k (m_{ijk})

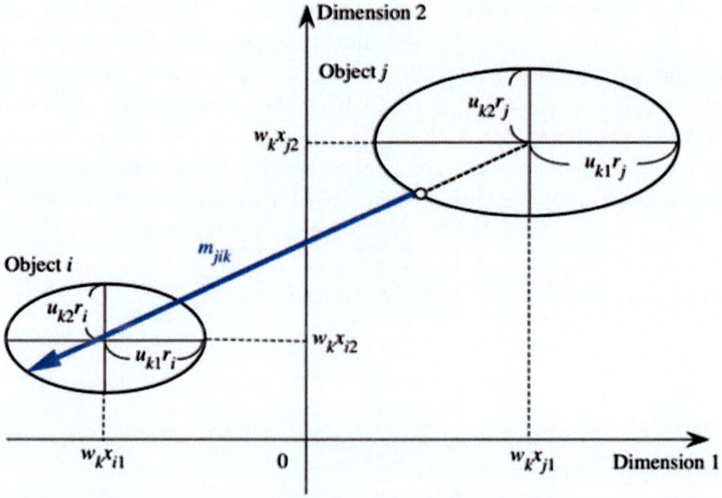

(b) Predicted value of the dissimilarity ω_{jik} from object j to object i for source k (m_{jik})

Fig. 5.5 Predicted values of ω_{ijk}(a) and ω_{jik} (b)

Figure 5.5 shows how m_{ijk} and m_{jik}, the predicted values of ω_{ijk} and ω_{jik}, are represented in the two-dimensional configuration of objects for source k where object i and object j are represented as points and ellipses centred at those points in the object configuration for source k. Object i is represented by a point $(w_k x_{i1}, w_k x_{i2})$ and an ellipse which has the axes of lengths $2u_{k1}r_i$ and $2u_{k2}r_i$ along dimensions 1 and 2, respectively. Analogously, object j is represented by a point $\left(w_k x_{j1}, w_k x_{j2}\right)$ and an ellipse with axes of lengths $2u_{k1}r_j$ and $2u_{k2}r_j$ along the two dimensions. This says that the orientation of the dimensions of the common object configuration **X** is uniquely determined up to transpositions and reflections (see Sect. 5.4; Carroll and Chang [7] and Okada and Imaizumi [14]).

The term m_{ijk} is represented by the length of the red arrow emitting from the small open circle on the ellipse representing object i in Fig. 5.5a and, similarly, m_{jik} is represented by the length of the blue arrow emitting from the small open circle on the ellipse representing object j in Fig. 5.5b. In Fig. 5.5a, the length of the dashed line inside the ellipse of object i depicts $v_{ijk}r_i$ and, similarly, in Fig. 5.5b, the length of the dashed line inside the ellipse of object j shows $v_{jik}r_j$. The lengths of the red and the blue arrowed lines representing m_{ijk} in Fig. 5.5a and m_{jik} in Fig. 5.5b, respectively, are not equal: thus they represent the asymmetry between object i and object j.

More general models [17–20] and a restricted model [16] have been developed based on the present model.

The algorithm to derive the common object configuration **X** ($x_{is}; i = 1, 2, \ldots, n, s = 1, 2, \ldots, r$) which is common to all sources, the asymmetry weight configuration **U** ($u_{ks}; k = 1, 2, \ldots, K, s = 1, 2, \ldots, r$), and the symmetry weights w_k ($k = 1, 2, \ldots, K$) from the observed two-mode three-way proximities ω_{ijk} $(i, j[i \neq j] = 1, 2, \ldots, n, k = 1, 2, \ldots, K)$ is the extension of the algorithm for one-mode two-way proximities (see Sect. 3.3.1; Okada and Imaizumi [12]). The badness-of-fit measure of the model for source k is defined by

$$S_k = \sqrt{\frac{\sum_{i=1}^{n}\sum_{\substack{j=1\\ i\neq j}}^{n}\left(m_{ijk} - \hat{m}_{ijk}\right)^2}{\sum_{i=1}^{n}\sum_{\substack{j=1\\ i\neq j}}^{n}\left(m_{ijk} - \bar{m}_k\right)^2}}, \tag{5.15}$$

where $\hat{m}_{ijk}$ is the disparity of m_{ijk} [21] which satisfies the monotonic relationship with d_{ijk} as follows

if $\omega_{ijk} < \omega_{h\ell k}$, then $\hat{m}_{ijk} \leq \hat{m}_{h\ell k}$ (when ω_{ijk} is dissimilarity)
and
if $\omega_{ijk} < \omega_{h\ell k}$, then $\hat{m}_{ijk} \geq \hat{m}_{h\ell k}$ (when ω_{ijk} is similarity).

Moreover, $\bar{m}_k$ is the mean of m_{ijk} for source k.

$$\bar{m}_k = \frac{\sum_{\substack{i=1 \\ i \neq j}}^{n} \sum_{j=1}^{n} m_{ijk}}{n(n-1)} \tag{5.16}$$

The badness-of-fit measure of Eq. (5.15) is the Stress [21, 22] based on the Stress formula II [23] for source k. The badness-of-fit measure of the model for all the K sources is defined as

$$S = \sqrt{\frac{1}{K} \sum_{k=1}^{K} S_k^2}. \tag{5.17}$$

The algorithm to derive the object configuration $\mathbf{X}$ and the radius r_i $(i = 1, 2, \ldots, n)$ which are common to all sources, the asymmetry weight configuration $\mathbf{U}$, and the symmetry weight w_k $(k = 1, 2, \ldots, K)$ minimizing the Stress S in (5.17) is the two-mode three-way extension of the algorithm for the one-mode two-way data (see Sect. 3.3.1; Okada and Imaizumi [12]) and it is described in detail by Okada and Imaizumi [14].

5.2.2.3 Slide Vector Model

The slide vector model [9–11] is formulated in terms of squared proximities (see Sect. 2.3.2.3);

$$\omega_{ij}^2 = d_{ij}^2 + \sum_{s=1}^{r} z_s^2 + 2 \sum_{s=1}^{r} z_s \left(x_{is} - x_{js}\right) + \varepsilon_{ij}, \tag{5.18}$$

where z_s is the s-th element of the slide vector $\mathbf{z} = (z_1, z_2, \ldots, z_s)'$, and d_{ij} is the Euclidean distance between the two points representing objects i and j in the object configuration, namely $d_{ij} = \sqrt{\sum_{s=1}^{r} \left(x_{is} - x_{js}\right)^2}$, where x_{is} is the coordinate of object i along dimension s in the configuration. Thus, the observed proximity ω_{ij} can be modelled as follows

$$\omega_{ij} = \sqrt{\sum_{s=1}^{r} \left(x_{is} - x_{js} + z_s\right)^2} + \varepsilon_{ij}. \tag{5.19}$$

The two-mode three-way extension of the slide vector model [9] is represented by

$$\omega_{ijk} = \sqrt{\sum_{s=1}^{r} w_{ks} \left(x_{is} - x_{js} + z_s\right)^2} + \varepsilon_{ijk}, \tag{5.20}$$

where $w_{ks}^{1/2}$ is the weight for source k along dimension s, which has the same effect of the weight in the INDSCAL model for expanding or shrinking both the slide vector and the dimension s of the configuration of points representing the objects. How to interpret the resulting configuration of objects is similar to the slide vector model for one-mode two-way data (see Sect. 2.3.2.3). How to interpret the weight w_{ks} is similar to INDSCAL model (see Sect. 5.2.1).

5.2.2.4 PREFSCAL

The aim of PREFSCAL is to analyse two-mode two-way data [24] by a model of unfolding in order to investigate the relationships between two sets of objects [25, 26]. In one-mode two-way data, the two ways usually originate from one (repeated) mode. But, when we assume that such two ways originate from two different modes, the one-mode two-way data can be treated as two-mode two-way data. Then, we can deal with $(n \times n)$ one-mode two-way data as $(n \times n)$ two-mode two-way data (see Sect. 5.1). Analogously, $(n \times n \times K)$ two-mode three-way data can be regarded as $(n \times n \times K)$ three-mode three-way data by assuming that the first and the second ways originate from two different modes. The three-way extension of PREFSCAL [27] can analyse $(n \times n \times K)$ two-mode three-way data as $(n \times n \times K)$ three-mode three-way data where ω_{ijk} is not necessarily equal to ω_{jik} for $k = 1, 2, \ldots, K$.

The PREFSCAL model can be regarded as a generalized slide vector model where each object has its own slide vector. The three-way extension of the unfolding model and the PREFSCAL algorithm are briefly described in Sect. 2.3.2.2.

5.2.3 *Representing Asymmetry by Relationships Among Dimensions*

The present subsection describes the method which represents the asymmetric relationships among objects indirectly by means of the asymmetric relationships among dimensions and not directly among the objects themselves. It is supposed that each dimension represents different underlying types of objects, which represent fundamental aspects of objects similarly to factors in factor analysis. The asymmetric relationships among objects are accounted for by (a) the asymmetric relationships among dimensions or types and (b) the weights of the dimensions for each object which represent the saliences of each type for source k.

DEDICOM [28, 29] was first introduced to analyse one-mode two-way data (see Sect. 3.3.2.1), and the DEDICOM model for two-mode three-way data is given by

$$\mathbf{\Omega}_k \cong \mathbf{X}\mathbf{D}_k\mathbf{R}\mathbf{D}_k\mathbf{X}', \ (k = 1, 2, \ldots, K), \tag{5.21}$$

where $\mathbf{\Omega}_k$ is the $n \times n$ matrix of the asymmetric relationships among the n objects for source k, $\mathbf{X}$ is the $(n \times r)$ matrix where the element (i, s) is the salience or the significance of dimension or type s for object i, $\mathbf{D}_k$ is the $(r \times r)$ diagonal matrix whose s-th diagonal element is the weight of dimension or type s for source k, and $\mathbf{R}$ is the $(r \times r)$ matrix which represents the asymmetric relationships among the r dimensions or types with entry (s, t) not necessarily equal to (t, s).

Equation (5.21) is similar to Eq. (3.21) which represents the DEDICOM model of one-mode two-way data. In Eq. (5.21), $\mathbf{X}$ can be multiplied by $\mathbf{D}_k$ which results in $\mathbf{XD}_k$, or $\mathbf{R}$ can be multiplied by $\mathbf{D}_k$ from the left and the right which results in $\mathbf{D}_k\mathbf{RD}_k$. In the first case, when we consider $\mathbf{XD}_k$ as the configuration of objects for source k, Eq. (5.21) is equivalent to Eq. (3.21). In this case $\mathbf{R}$, which conveys the asymmetric relationships among dimensions or types, is common to all sources, but each source has its own idiosyncratic configuration of objects $\mathbf{XD}_k$ whose dimensions were extended or shrunk by $\mathbf{D}_k$ to represent the differences among sources, which is similar to $\mathbf{W}_k^{1/2}$ in the INDSCAL model. On the other hand, when we consider $\mathbf{D}_k\mathbf{RD}_k$ for the asymmetric relationships among dimensions for source k, each source has its own idiosyncratic asymmetric relationships among dimensions which represents the differences among sources, while all sources have the same object configuration $\mathbf{X}$. In order to derive $\mathbf{X}$, $\mathbf{R}$, and $\mathbf{D}_k$ to analyse $\mathbf{\Omega}_k (k = 1, 2, \ldots, K)$, Harshman [28] suggests the NILES algorithm of the regression procedure [30].

Unlike the model which represents the asymmetry by relationships among objects, we cannot directly grasp the asymmetric relationships among objects by the model which represents the asymmetry by relationships among dimensions. This is not only a drawback but has also the merit that the asymmetric relationships among dimensions or types are revealed. The dimensions or types are a set of underlying components for representing asymmetric relationships among objects. This makes it possible to comprehend asymmetric relationships among objects by the small number of dimensions or types which are much less than the number of objects. Harshman et al. [31] describe how to interpret the asymmetric relationships of the result of word association derived by DEDICOM in detail.

5.3 One-Mode Three-Way Asymmetric Methods

As described in Sect. 5.2.2.3, the slide vector model (see Sect. 2.3.2.3; Zielman and Heiser [9–11]), originally developed to analyse one-mode two-way data, was extended to analyse two-mode three-way data.

As a second extension of the slide vector model to deal with three-way data, De Rooij and Heiser [3] proposed an extension to analyse one-mode three-way data where two types of asymmetry are assumed: the asymmetry between the first and the second way, and the asymmetry between the second and the third way. Each of them is represented by a slide vector, so that the resulting configuration consists of points representing objects and two slide vectors. Figure 5.6 shows a configuration of three objects i, j and k, and two slide vectors: $\mathbf{z} = (z_1, z_2, \ldots, z_r)'$ which represents

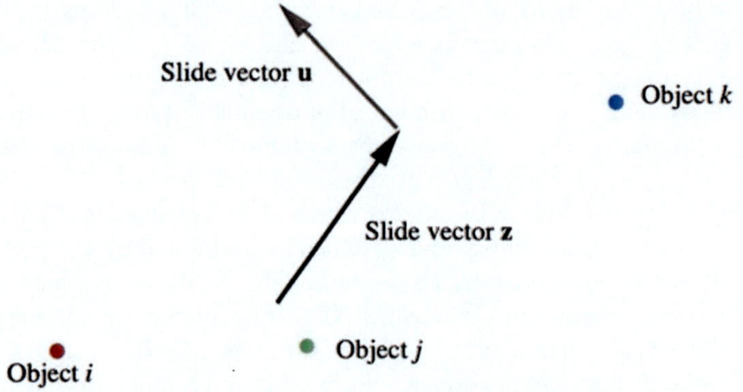

Fig. 5.6 Two-dimensional configuration of the one-mode three-way extension of the slide vector model. The configuration represents three objects i, j and k, and two slide vectors $\mathbf{z}$ and $\mathbf{u}$

the asymmetry between the first and the second way, and $\mathbf{u} = (u_1, u_2, \ldots, u_r)'$ to represent the asymmetry between the second and the third way.

Figure 5.7 illustrates how the two-dimensional one-mode three-way slide vector model represents two kinds of asymmetries by $\mathbf{z}$ and $\mathbf{u}$. The figure shows three objects i, j and k in a two-dimensional space, where each object is represented by a set of three points. For example, objects $i^{(1)}$, $i^{(2)}$ and $i^{(3)}$ in Fig. 5.7 represent object i on the first, second and third ways, respectively. Note that objects $i^{(1)}, j^{(1)}$ and $k^{(1)}$ correspond to objects i, j and k in Fig. 5.6, respectively.

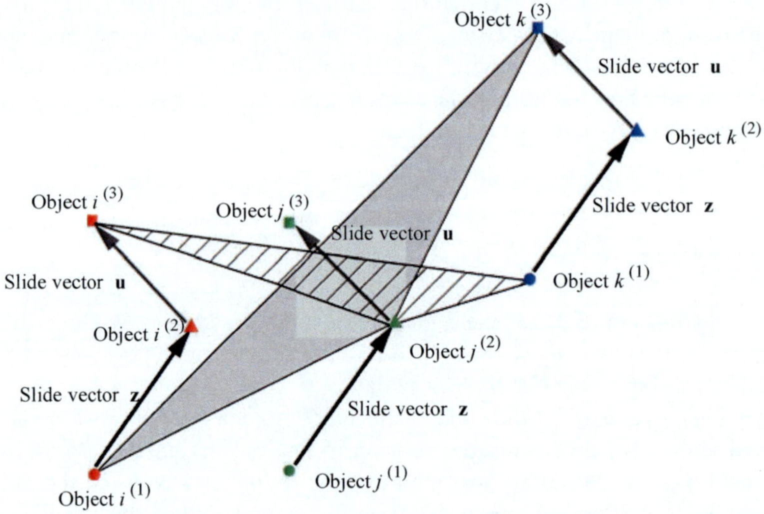

Fig. 5.7 Predicted values of ω_{ijk} and ω_{jik} in the two-dimensional model of one-mode three-way extension of the slide vector model

Object i is shifted by the slide vector **z** from object $i^{(1)}$ to object $i^{(2)}$, and then it is shifted again from object $i^{(2)}$ to object $i^{(3)}$ by the slide vector **u**. Let $x_{i^{(1)}s}$ (=x_{is}) be the coordinate of object i of the first way along dimension s. The coordinate of object i of the second way (object $i^{(2)}$) along dimension s is

$$x_{i^{(2)}s} = x_{i^{(1)}s} + z_s \tag{5.22}$$

where z_s is the coordinate of the slide vector **z** along dimension s. Then, the coordinate of object i of the third way (object $i^{(3)}$) along dimension s is

$$x_{i^{(3)}s} = x_{i^{(2)}s} + u_s = x_{i^{(1)}s} + z_s + u_s \tag{5.23}$$

where u_s is the coordinate of the slide vector **u** along dimension s. The two slide vectors can be equal (**z** = **u**) as in the less general model of [3].

Let d_{ijk} be the three-way distance of the three objects i, j and k in the configuration. The predicted value of ω_{ijk} is given by d_{ijk} which is defined as the square root of the sum of the three squared interpoint distances [3–5, 32–34]

$$\omega_{ijk} = d_{ijk} + \varepsilon_{ijk}, \tag{5.24}$$

where

$$d_{ijk} = \sqrt{\sum_{s=1}^{r}\left(x_{i^{(1)}s} - x_{j^{(2)}s}\right)^2 + \sum_{s=1}^{r}(x_{i^{(1)}s} - x_{k^{(3)}s})^2 + \sum_{s=1}^{r}\left(x_{j^{(2)}s} - x_{k^{(3)}s}\right)^2}, \tag{5.25}$$

or

$$d_{ijk} = \sqrt{\sum_{s=1}^{r}\left(x_{is} - x_{js} - z_s\right)^2 + \sum_{s=1}^{r}(x_{is} - x_{ks} - z_s - u_s)^2 + \sum_{s=1}^{r}\left(x_{js} - x_{ks} - u_s\right)^2}. \tag{5.26}$$

The shaded triangle in Fig. 5.7 illustrates what d_{ijk} means. The first term under the square root in Eqs. (5.25) and (5.26) is the squared distance between objects $i^{(1)}$ and $j^{(2)}$. The distance is displayed in Fig. 5.7 by the length of the side connecting objects $i^{(1)}$ and $j^{(2)}$ of the shaded triangle formed by the three points representing objects $i^{(1)}, j^{(2)}$ and $k^{(3)}$. The second term is the squared distance between objects $i^{(1)}$ and $k^{(3)}$ which is shown by the squared length of the side connecting objects $i^{(1)}$ and $k^{(3)}$ of the shaded triangle. The third term is the squared distance between objects $j^{(2)}$ and $k^{(3)}$ which corresponds to the squared length of the side connecting objects $j^{(2)}$ and $k^{(3)}$ of the shaded triangle.

The striped triangle in Fig. 5.7 formed by the three points representing objects $k^{(1)}, j^{(2)}$ and $i^{(3)}$ depicts d_{kji} which is the predicted value of ω_{kji} as follows.

$$d_{kji}=\sqrt{\sum_{s=1}^{r}\left(x_{k^{(1)}s}-x_{j^{(2)}s}\right)^2+\sum_{s=1}^{r}(x_{k^{(1)}s}-x_{i^{(3)}s})^2+\sum_{s=1}^{r}\left(x_{j^{(2)}s}-x_{i^{(3)}s}\right)^2}. \quad (5.27)$$

Hence, the squared lengths of the three sides connecting the pairs of objects ($k^{(1)}$ and $j^{(2)}$), ($k^{(1)}$ and $i^{(3)}$) and ($j^{(2)}$ and $i^{(3)}$) correspond to the three terms in (5.27), respectively.

The algorithm to derive the object configuration $\mathbf{X} = [x_{is}]$ ($i = 1, 2, \ldots, n$; $s = 1, 2, \ldots, r$) and the two slide vectors $\mathbf{z}$ and $\mathbf{u}$ was developed for minimizing the loss function

$$F = \sum_{i=1}^{n}\sum_{j=1}^{n}\sum_{k=1}^{n}\varphi_{ijk}\left(\omega_{ijk} - d_{ijk}\right)^2, \quad (5.28)$$

where φ_{ijk} is the weight to code possible missing values. Although the model fits the relationships among triples of objects, the three-way distance d_{ijk} is actually defined by Eqs. (5.25), (5.26) or (5.27), so that the loss function depends on the squared distances between pairs of objects or on the pairwise relationships. The detailed algorithm which utilizes the SMACOF algorithm is presented in De Rooij and Heiser [3].

5.4 Applications

In the present section, an application of the two-mode three-way asymmetric multidimensional scaling presented in Okada and Imaizumi [14] is shown with discussions from different points of view not mentioned in their earlier study. The application deals with the analysis of the intergenerational occupational mobility among eight occupational categories. Original data [35, pp. 46–47, Table 2.12] which consist of the intergenerational occupational mobility from fathers to sons among eight occupational categories in four years, are shown in Table 1.6. The details of the data are reported in Sect. 1.3. The eight occupational categories are

Professional occupations
Non-manual occupations employed by **large** enterprises
Non-manual occupations employed by **small** enterprises
Non-manual self-employed occupations
Manual occupations employed by **large** enterprises
Manual occupations employed by **small** enterprises
Manual self-employed occupations, and

Farm occupations.

As mentioned earlier in Sect. 1.3, the emboldened word(s) in each occupational category are used to stand for the occupational category henceforth.

The four tables in Table 1.6 contain the two-mode three-way proximities among the eight occupational categories over four years (8 occupational categories $\times$ 8 occupational categories $\times$ 4 years). In each table, it is clear that differences among the eight row marginal totals emerge, and differences among the eight column marginal totals as well. For example, in the year 1955 the row marginal total of the non-manual small occupational category is 32, while the row marginal total of the farm occupational category is 1091. Moreover, the column marginal total of the non-manual small occupational category is 94, while the column marginal total of the farm occupational category is 769. The marginal total of row j represents the whole of the outflow from occupational category j, which reflects its unattractiveness, while the marginal total of column j represents the whole of the inflow to occupational category j, which reflects its attractiveness. The sum of the row and column marginal totals of category j represents the amount of manpower of such an occupational category and the differences of such sums highlight the differences of the manpower among occupational categories. Whether or not occupational categories are attractive is one of the factors which influence the intergenerational occupational mobility. But it is desirable to remove the effect of the differences in the manpower among the occupational categories to reveal the structure of the intergenerational occupational mobility among occupational categories [14, 15, 36, 37]. The analysis of the marginal differences can often reveal interesting aspects of the asymmetry of the relationships [38].

Each table was rescaled by multiplying with a rescaling constant to remove the differences in the manpower among the occupational categories. The i-th row and the i-th column of the k-th table were multiplied with a rescaling constant c_{ki}. Let s_{ijk} be the similarity from occupational category i (of sons) to occupational category j (of fathers) at k-th table. The rescaling constant c_{ki} is determined to satisfy

$$c_{ki}\left(\sum_{j=1}^{n}(s_{ijk}+s_{jik})\right)=\frac{1}{n}\sum_{i=1}^{n}\left\{c_{ki}\left(\sum_{j=1}^{n}(s_{ijk}+s_{jik})\right)\right\},\ (\text{for } k=1,2,\ldots,K) \tag{5.29}$$

so that the sum of the row i and column i marginal totals is equal to the mean of the eight sums of the row and the column marginal totals in the rescaled k-th table for all eight occupational categories [15, p. 283; 31, 36]. The right-hand side of Eq. (5.29) is different from one source to another. But the badness-of-fit measure defined by Eq. (5.17) is source conditional [39, 40, pp. 60–63], and the differences among the right-hand side of Eq. (5.29) do not affect the result.

The rescaled data are reported in Table 5.1. In the first table of Table 5.1, which represents the rescaled intergenerational occupational mobility in 1955, the summation of the cells in the first row (professional occupational category) is 221.34, while

Table 5.1 Rescaled intergenerational occupational mobility

Father's occupation	Son's occupation year 1955							
	1	2	3	4	5	6	7	8
1 Professional	118.22	33.85	39.30	3.38	3.14	8.64	9.09	5.71
2 Non-manual large	39.49	53.70	39.38	22.00	15.75	17.32	7.59	10.81
3 Non-manual small	4.91	29.54	42.87	23.58	16.46	30.17	7.94	0.00
4 Non-manual self	35.47	50.77	58.96	72.97	20.75	22.48	23.64	5.33
5 Manual large	9.43	9.45	54.86	5.66	73.71	38.61	11.85	3.54
6 Manual small	5.76	14.44	15.09	10.37	45.05	64.89	20.17	7.15
7 Manual self	13.64	36.45	37.03	16.37	42.31	65.16	71.81	9.23
8 Farm	23.49	37.52	28.80	27.04	47.49	41.58	27.68	98.19
Father's occupation	Son's occupation year 1965							
	1	2	3	4	5	6	7	8
1 Professional	102.92	32.94	26.39	3.85	14.19	19.78	11.57	1.59
2 Non-manual large	45.29	50.72	31.66	18.34	22.24	13.85	9.72	4.44
3 Non-manual small	22.62	23.22	54.12	13.83	26.19	17.75	5.93	0.81
4 Non-manual self	28.89	44.23	41.48	94.89	20.82	25.92	17.18	3.74
5 Manual large	17.02	31.78	34.91	1.49	56.94	41.99	10.42	3.06
6 Manual small	2.47	12.46	32.96	6.48	34.36	53.23	15.57	5.87
7 Manual self	7.71	42.12	43.50	30.32	37.21	50.59	78.91	4.99
8 Farm	34.87	41.28	45.52	28.66	65.46	88.51	30.35	57.93
Father's occupation	Son's occupation year 1975							
	1	2	3	4	5	6	7	8
1 Professional	123.30	44.28	52.80	15.61	13.54	10.97	7.35	3.75
2 Non-manual large	37.47	76.59	46.35	27.50	24.68	22.21	11.17	4.10
3 Non-manual small	14.67	30.31	49.13	11.44	17.00	34.44	1.92	2.36
4 Non-manual self	34.33	55.00	71.88	89.51	36.18	32.57	38.90	3.34
5 Manual large	29.78	29.62	34.01	12.06	75.85	54.73	10.66	3.62
6 Manual small	12.79	16.66	36.35	7.12	40.61	66.74	11.99	2.45
7 Manual self	27.58	36.87	55.81	28.67	40.85	52.75	106.13	6.89
8 Farm	32.99	45.11	76.92	30.90	85.47	115.40	40.83	65.20

(continued)

Table 5.1 (continued)

Father's occupation	Son's occupation year 1985							
	1	2	3	4	5	6	7	8
1 Professional	88.45	37.35	49.02	7.38	6.72	11.45	1.53	0.00
2 Non-manual large	39.93	63.03	42.83	17.18	24.78	14.83	7.15	1.58
3 Non-manual small	11.14	23.36	47.15	8.92	11.28	28.22	6.18	0.00
4 Non-manual self	39.82	32.65	57.97	97.40	25.39	22.07	18.42	2.41
5 Manual large	42.54	32.61	36.10	4.48	72.56	27.07	12.43	0.00
6 Manual small	14.00	23.72	38.48	7.64	32.22	52.75	15.89	3.64
7 Manual self	18.41	34.85	37.10	24.56	31.07	57.40	95.79	3.77
8 Farm	44.31	41.62	55.59	36.80	68.68	98.38	40.16	51.78

in the first column (professional occupational category) is 250.41, and their sum is 221.34 + 250.41 = 471.75. For the eighth row and column (farm occupational category), the marginal sums are 331.79 and 139.96, respectively, and their sum is 471.75. Another feature of such a rescaling is that the ratio of two cells, (i, j) and (j, i), is not altered before and after the rescaling [cf. Borg and Groenen [41, p. 504], Zielman and Heiser [10], Groenen and Heiser [42], see Eq. (1.6)], because both cells (i, j) and (j, i) of the k-th table are multiplied by $c_{ki} \cdot c_{kj}$. The values in Table 5.1 are similarities: the larger value represents the larger mobility from the son's occupational category (row) to the fathers' occupational category (column). High values will lead to small distances; low values will be represented by large distances. The two-mode three-way proximities shown in Table 5.1 were analysed by the two-mode three-way asymmetric multidimensional scaling of Okada and Imaizumi [14]. As Eq. (5.15) shows, the diagonal elements of each table in Table 5.1 are not dealt with in the algorithm to derive the result.

The process of the analysis was as follows [14]. Five series of analyses were done. The first series of analyses consists of eight analyses. The first series starts in the eight-dimensional space using the rational initial configuration. Then, the analysis in the seven-dimensional space using the seven-dimensional "principal components" of the eight-dimensional results. The analysis in the six-dimensional space using the six-dimensional "principal components" of the seven-dimensional results. ... The analysis in the unidimensional space using the unidimensional "principal components" of the two-dimensional results. The first series of analyses results in eight minimized S for the eight- through unidimensional spaces. The second series consists of seven analyses: the first analysis in the seven-dimensional space, the second analysis in the six-dimensional space, ..., the seventh analysis in the unidimensional space. The second series of analyses results in seven minimized S for the seven- through unidimensional spaces. The third series of analyses results in six minimized S for the six- through unidimensional spaces. ... The fifth series of analyses results in four minimized S for the four- through unidimensional spaces. The five series of analyses gave one minimized S in eight-dimensional space, two minimized S in seven-dimensional space, ..., and, five minimized S in four-, three-, two- and unidimensional spaces. For each dimensional space, the minimum of the minimized S was chosen. The values of the minimized S obtained for five- through unidimensional spaces were 0.335, 0.335, 0.349, 0.378, and 0.453, respectively. The two-dimensional result was chosen as the solution, based on both the value of S with respect to the dimensionality and the interpretation of the configuration. Figure 5.8 illustrates the common object configuration of eight occupational categories of the obtained solution in the two-dimensional optimal space. The coordinates of the occupational categories and the radii are reported in Table 5.2, while both the asymmetry and symmetry weights are shown in Table 5.3.

As mentioned earlier (Sect. 5.2.2.2), the orientation of the dimensions of the common object configuration $\mathbf{X}$ is uniquely determined up to transpositions and reflections. This aspect will be examined later.

In the common object configuration shown in Fig. 5.8, each occupational category is represented by a point (diamond) and a circle centred at that point. Since the radius of the circle representing the non-manual small occupational category is zero, it is represented only by a diamond (without any circle).

The asymmetry weight configuration is displayed in Fig. 5.9, where each year is represented by a diamond marker.

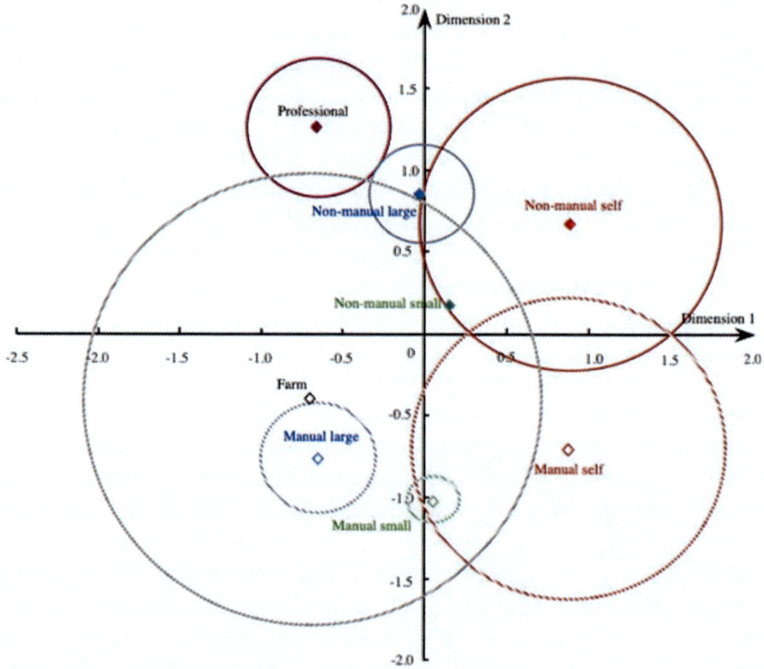

Fig. 5.8 Common object configuration of the eight occupational categories. *Note* The non-manual occupational category is represented by a solid diamond and circle, while the manual one is represented by an open diamond and a dotted circle. The self-employed category is in red, small enterprises in green, large enterprises in blue, professional category in violet, while farm category is black.

Table 5.2 Configuration of the occupational categories

Occupational category	Dim 1	Dim 2	Radius
1 Professional	−0.648	1.256	0.428
2 Non-manual large	−0.020	0.853	0.303
3 Non-manual small	0.160	0.171	0.000
4 Non-manual self	0.896	0.666	0.913
5 Manual large	−0.640	−0.773	0.373
6 Manual small	0.064	−1.030	0.149
7 Manual self	0.880	−0.717	0.961
8 Farm	−0.691	−0.407	1.393

Table 5.3 Asymmetry and symmetry weights

Year	Asymmetry weight		Symmetry weight
	Dim 1	Dim 2	
1955	0.424	0.379	0.503
1965	0.454	0.446	0.428
1975	0.497	0.438	0.402
1985	0.496	0.415	0.427

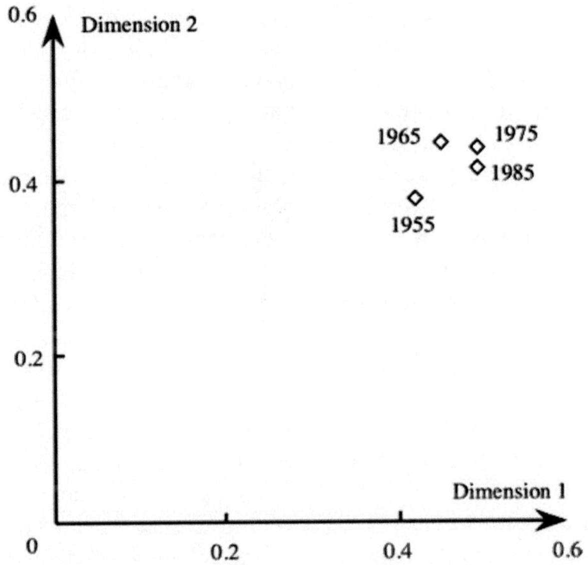

Fig. 5.9 Asymmetry weight configuration of the four years

It is necessary to point out that, similarly to the INDSCAL model [7], the orientation of the dimensions of the common object configuration in Fig. 5.8 is uniquely determined up to reflection and transposition, and rotation is not allowed. The uniqueness of the orientation of the dimensions of the common object configuration is due to the asymmetry weights. When the asymmetry weights in Fig. 5.9 are aligned on the bisector line (which means that the asymmetry weights along dimension 1 and dimension 2 are equal for each year), the uniqueness of the orientation of the dimensions does not hold, similarly to the INDSCAL model [8, p. 25].

Thus, the dimensions of the common object configuration have the uniqueness of the orientation in general, but if the asymmetry weights along all dimensions are equal for all sources (years), such a uniqueness of the orientation is not given. Even if the four points of the asymmetry weight configuration in Fig. 5.9 are not perfectly aligned on the bisector line, they are not far from this situation.

In order to evaluate the uniqueness of the common object configuration, Okada and Imaizumi [14, p. 220] performed two supplementary analyses as follows. Two asymmetry weight configurations were derived by rotating the dimensions of the two-dimensional asymmetry weight configuration of the optimal solution (Fig. 5.9) 10° clockwise and counterclockwise. For each of the rotated asymmetry weight configurations, the common object configuration (configuration of the occupational categories and radii) and the symmetry weights minimizing S were derived in a two-dimensional space. The resulting S values were 0.448 and 0.458 for the clockwise and counterclockwise cases, respectively. These values are substantially larger than the optimal S (0.378) of the solution which implies that the dimensions of the common object configuration do not have the indeterminacy of the orientation.

The dimensions of Fig. 5.8 can be interpreted very clearly. Dimension 1 of the solution differentiates four occupational categories (from right to left); self-employed, employees of small enterprises, employees of large enterprises, and professional or farm occupational categories. Dimension 2 represents the difference between non-manual occupational categories, which are in the upper area of the configuration (non-manual self, non-manual small, non-manual large and professional occupational categories), versus manual categories, which are in the lower area of the configuration (manual self, manual small, manual large and farm occupational categories).

The sum of the squared coordinates of the eight occupational categories along dimensions 1 and 2 is 2.914 and 5.116, respectively, and their sum is 8.030 (very close to 8 which is the number of the occupational categories) because of the normalization of the common object configuration, where the origin is the centroid [14], Eq. (3.4)). Moreover, the two sums of squares show that the variability of the coordinates along dimension 2 is larger than along dimension 1. The larger variability of dimension 2 suggests that dimension 2 has more variation than dimension 1, and that the configuration of the occupational categories is vertically longer. This implies that the mobility *within* both non-manual and manual categories is larger than the mobility *between* non-manual and manual categories. This is also confirmed by Table 5.4 which shows the average $d_{ij} - r_i + r_j$ within and between non-manual and manual categories in the common object configuration of Fig. 5.8.

Table. 5.4 Average $d_{ij} - r_i + r_j$ within/between non-manual and manual occupational categories in the common object configuration

Father's occupation	Son's occupation	
	Non-manual	Manual
Non-manual	1.047	2.017
Manual	1.401	1.016

Note The value in cell (1,1) is the average ($d_{ij} - r_i + r_j$) within non-manual categories (excluding diagonal cells), the value in cell (1,2) is the average from non-manual to manual categories, the value in cell (2,1) is the average from manual to non-manual categories and, finally, the value in cell (2,2) is the average within manual categories (excluding diagonal cells)

Both values in cells (1,1) and (2,2) are smaller than values in cells (1,2) and (2,1) which again suggests that mobility within both non-manual and manual occupational categories is larger than mobility between non-manual and manual occupational categories.

The four subtables of Table 5.5 report the rescaled intergeneration mobility of Table 5.1 averaged for both non-manual and manual occupational categories for the four years. In all subtables of Table 5.5, the diagonal values which represent the mobility *within* both non-manual (cell (1,1)) and manual (cell (2,2)) categories are larger than the off-diagonal values which represent the mobility *between* non-manual and manual categories (cells (1,2) and (2,1)). The derived common object configuration and the resulting $d_{ij} - r_i + r_j$ are consistent with such four subtables in Table 5.5.

Figure 5.8 and Table 5.2 show that the radius of the non-manual category is smaller than the radius of the manual category, i.e. 0.411 and 0.719, respectively.

This suggests that the intergenerational occupational mobility from manual to non-manual categories was larger than the reverse. In Table 5.4, the average $d_{ij} - r_i + r_j$ from non-manual to manual categories (2.017 in cell (2,1)) is larger than the average $d_{ij} - r_i + r_j$ from manual to non-manual categories (1.401 in cell (2,1)) which confirms that the intergenerational occupational mobility from manual to non-manual occupational categories was larger than the reverse.

In Fig. 5.8, the circle with the largest radius corresponds to the farm occupational category. Generally speaking, an occupational category with a large radius implies that the outflow from that category to the others tends to be large, and, conversely, that the inflow directed to that category tends to be small. More specifically, the larger the radius of an occupational category, the greater the tendency of the sons to leave the occupational category where their fathers are or were occupied, and the smaller the tendency of the sons to move into the occupational category where their fathers are or were occupied. This is consistent with the fact that the farm occupational category has the largest row sum and the smallest column sum in all four tables of Table 5.1, and it is also compatible with the general very large outflow of manpower from the farm category [35, p. 48).

The non-manual small occupational category presents the circle with the smallest radius which is equal to zero. Differently from the one-mode two-way radius-distance

Table 5.5 Average rescaled intergenerational occupational mobility within/between non-manual and manual occupational categories

1955	Son's occupation	
Father's occupation	Non-manual	Manual
Non-manual	31.72	12.80
Manual	21.59	29.98
1965	Son's occupation	
Father's occupation	Non-manual	Manual
Non-manual	27.73	13.48
Manual	25.85	32.36
1975	Son's occupation	
Father's occupation	Non-manual	Manual
Non-manual	36.80	16.53
Manual	32.08	38.85
1985	Son's occupation	
Father's occupation	Non-manual	Manual
Non-manual	22.97	15.17
Manual	30.80	32.56

Note. For each year, the value in cell (1,1) is the average of the within non-manual categories values (excluding diagonal cells) of the corresponding subtable of Table 5.1; the value in cell (1,2) is the average of the values from non-manual to manual categories of the corresponding subtable of Table 5.1; the value in cell (2,1) is the average of the values from manual to non-manual categories of the corresponding subtable of Table 5.1, and cell (2,2) is the average of the within manual categories values (excluding diagonal cells) of the corresponding subtable of Table 5.1.

model where $\min_i r_i = 0$ [12], the present model does not impose any condition on the minimum value of the radius. An occupational category with a small radius generally implies that the outflow tends to be small and, conversely, the inflow from the other categories tends to be large.

In the asymmetry weight configuration of Fig. 5.9 (Table 5.3), the asymmetry weights for dimension 1 are larger than those for dimension 2 for all four years which means that the asymmetry in the intergenerational occupational mobility along dimension 1 is larger than along dimension 2 for all four years. More concretely, the asymmetry in the intergenerational occupational mobility *within* both non-manual and manual occupational categories is larger (or more asymmetric) than *between* non-manual and manual occupational categories. This effect is stronger for 1975 and 1985.

For each table in Table 5.1, the absolute difference of cell (i, j) and cell (j, i), which represents the absolute asymmetry between categories i and j, was computed. Then, the average absolute asymmetry within and between non-manual and manual

Table 5.6 Average absolute asymmetry within/between non-manual and manual occupational categories

	Year			
Average absolute asymmetry	1955	1965	1975	1985
Within non-manual	12.18	8.60	13.97	13.07
Within manual	29.79	39.98	52.30	44.18
Between non-manual and manual	8.24	10.02	11.86	11.92

occupational categories were computed (excluding diagonal cells) for each table of Table 5.1 as reported in Table 5.6.

Again, the average absolute asymmetries are larger within than between the categories except for year 1965. The asymmetry weights reported in Table 5.3 agree with the average absolute asymmetries shown in Table 5.6.

The symmetry weights in the last column of Table 5.3 exhibit a decreasing trend from 1955 through 1975 and a slightly increase in 1985. This suggests that the symmetric intergenerational occupational mobility decreased from 1955 through 1975, and somewhat increased in 1985. The asymmetry weight along dimension 1 increased from 1955 through 1975, and decreased slightly in 1985. This suggests that the asymmetry of the intergenerational occupational mobility generally increased over the years with a slight decrease in 1985. The opposite trend can be observed looking at the asymmetry weights for both dimensions which is reasonable because the sum of the symmetric and asymmetric component is constant for all years [14, p. 209].

To validate the effectiveness of analysing the asymmetry in the present data, the symmetrized rescaled data of Table 5.1 were computed by summing the upper and lower triangular cells, and they were analysed by INDSCAL [7] which does not take into account the asymmetry in the data. The two-dimensional result was chosen as the solution (the maximized variances accounted for obtained in five- through unidimensional spaces were 0.963, 0.931. 0.877, 0.817 and 0.623, respectively). The common object configuration of the eight occupational categories and the weight configuration of the four years are displayed in Figs. 5.10 and 5.11, respectively.

Occupational categories are not represented as circles with different radii, but the configuration of points in Figs. 5.8 and 5.10 are very similar. The same interpretation of the dimensions is also possible for the two figures. Dimension 1 differentiates four occupational categories: self-employed, employees of small enterprises, employees of large enterprises and professional or farm occupational categories, while dimension 2 represents the difference between non-manual and manual occupational categories. Even though INDSCAL cannot represent the asymmetric relationships among occupational categories, it can catch the symmetric relationships which are very similar to the result obtained by the asymmetric multidimensional scaling.

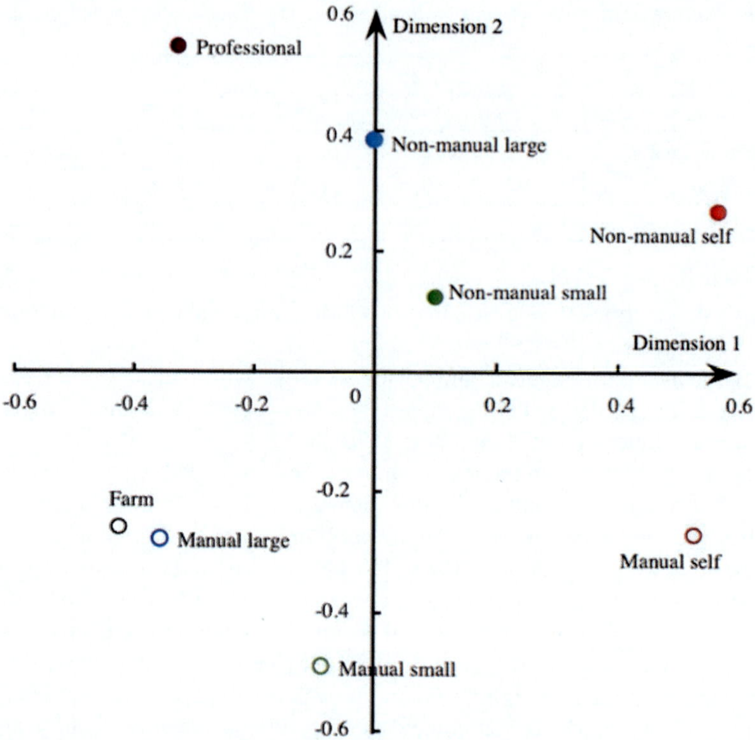

Fig. 5.10 Common object configuration of eight occupational categories derived from INDSCAL. *Note* The non-manual categories are represented by solid small circles, while the manual categories are represented by open small circle. The self-employed category is represented by a red circle, the employee of small enterprises is represented in green, the employee of large enterprises is represented in blue, the professional category is represented in violet, and the farm category is represented in black.

Figure 5.11 illustrates the weight configuration. Four points representing the four years are not aligned on the bisector line (see this section) which tells that the orientation of the dimensions of the common object configuration illustrated in Fig. 5.10 is uniquely determined up to reflection and transposition [8, p. 25]. No rotation of the dimensions of the configuration in Fig. 5.10 was done to match the dimensions of the configuration in Fig. 5.8.

The common object configuration of INDSCAL is normalized so that the origin is at the centroid and the sum of the squared coordinates is one along each dimension [7, p. 289]. This normalization is different from the one used for the common object configuration of Okada and Imaizumi [14] (see this section).

Each dimension of the common object configuration of Fig. 5.10 is multiplied with the square root of the weight represented in Fig. 5.11 [see Eq. (5.4)] to derive the configuration of the occupational categories for each year. For any year such a configuration results vertically longer than the common object configuration in

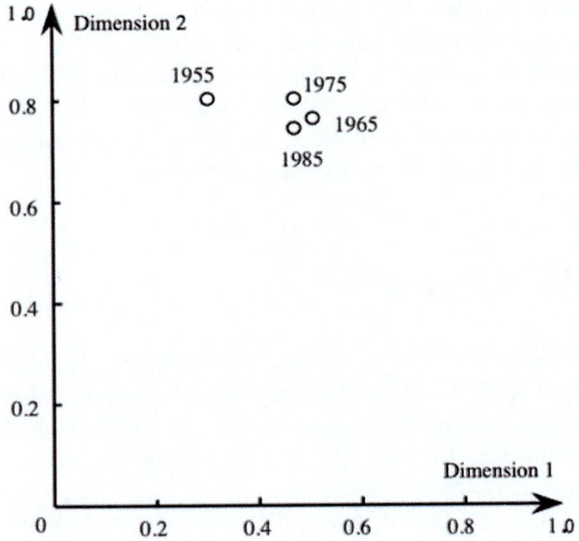

Fig. 5.11 Weight configuration of the four years derived from INDSCAL

Fig. 5.10, because the weight along dimension 2 is larger than the weight along dimension 1 for any years. Hence, dimension 2 is relatively more expanded than dimension 1, and distances along dimension 2 are relatively more stretched than distances along dimension 1, which means that the distance *between* non-manual and manual categories is relatively more stretched than the distance *within*. This tells that the intergenerational occupational mobility is larger within than between the non-manual and manual categories. This point is suggested by the results obtained by the asymmetric multidimensional scaling as well (see this section).

The results obtained by Okada and Imaizumi [14] disclosed other points, e.g. the intergenerational occupational mobility from non-manual to manual categories is smaller than the reverse, and the asymmetry is larger within than between non-manual and manual categories. The analysis by INDSCAL is not able to disclose these two points, because it does not consider the asymmetry in the intergenerational occupational mobility.

As describes earlier (this section), the present analysis did not consider the diagonal cells of each table [see Eq. (5.15)] which contain the number of sons who are in the same occupational category as their fathers. The diagonal cells present the largest element in most of the rows and columns of each table with three anomalies in 32 rows and nine anomalies in 32 columns, respectively. The diagonal values actually influence the intergenerational occupational mobility, but it is necessary to note here that this analysis focuses only on the mobility between different categories.

In the present section, several points not mentioned in the original paper have been discussed. The comparison of the results from the asymmetric multidimensional scaling with the results from INDSCAL [7] has been also performed. Such comparison shows that the results obtained from INDSCAL can represent only some of the aspects which the asymmetric analysis disclosed and such aspects are not related to the asymmetry of the intergenerational occupational mobility because INDSCAL neglects the asymmetry. The comparison of the results from the two models suggests that the asymmetric multidimensional scaling can account for the asymmetric relationships in the data effectively, and analysing asymmetry in the data is useful to reveal the underlying structure of the data.

5.5 Software

Two-mode three-way asymmetric similarities/dissimilarities can be analysed on the Website https://non-metric-mds.shinyapps.io/asymmetric_mds/ by the program for the two-mode three-way asymmetric multidimensional scaling [14] written in R and Shiny. How to use the program is briefly explained in the "Description" tab panel of the Shiny application above.

The maximum number of both objects and sources is 50.

On the Website, synthetic two-mode three-way asymmetric similarities/dissimilarities synthetic data are presented which can be used to test the program.

References

1. Carroll, J. D., & Arabie, P. (1980). Multidimensional scaling. *Annual Review of Psychology, 31*, 607–649.
2. Tucker, L. R. (1964). The extension of factor analysis to three-dimensional matrices. In N. Frederiksen, & H. Gulliksen (Eds.), *Contributions to mathematical psychology* (pp. 109–127). Holt Reinhart & Winston.
3. De Rooij, M., & Heiser, W. J. (2000). Triadic distance models for the analysis of asymmetric three-way proximity data. *British Journal of Mathematical and Statistical Psychology, 53*, 99–119.
4. Joly, S., & Le Calvé, G. (1995). Three-way distances. *Journal of Classification, 12*, 191–205.
5. Heiser, W. J., & Bennani, M. (1997). Triadic distance models: Axiomatization and least squares representation. *Journal of Mathematical Psychology, 41*, 189–206.
6. Tucker, L. R., & Messick, S. (1963). An individual difference model for multidimensional scaling. *Psychometrika, 28*, 333–367.
7. Carroll, J. D., & Chang, J. J. (1970). Analysis of individual differences in multidimensional scaling via an N-way generalization of "Eckart-Young" decomposition. *Psychometrika, 35*, 283–319.
8. Arabie, P., Carroll, J. D., & DeSarbo, W. S. (1987). *Three-way scaling and clustering*. Sage Publications.
9. Zielman, B., & Heiser, W. J. (1991). Analysis of asymmetry by a slide vector. *Research Report RR-91-05*. Department of Data Theory, Leiden University.
10. Zielman, B., & Heiser, W. J. (1993). Analysis of asymmetry by a slide-vector. *Psychometrika, 58*, 101–114.
11. Zielman, B., & Heiser, W. J. (1996). Models for asymmetric proximities. *British Journal of Mathematical and Statistical Psychology, 49*, 127–146.

12. Okada, A., & Imaizumi, T. (1987). Non-metric multidimensional scaling of asymmetric similarities. *Behaviormetrika, 21*, 81–96.
13. Zielman, B. (1991). Three-way scaling of asymmetric proximities. *Research Report RR-91-01*. Department of Data Theory, Leiden University.
14. Okada, A., & Imaizumi, T. (1997). Asymmetric multidimensional scaling of two-mode three-way proximities. *Journal of Classification, 14*, 195–224.
15. Okada, A. (1988). Asymmetric multidimensional scaling of car switching data. In W. Gaul & M. Schader (Eds.), *Data, expert knowledge and decisions* (pp. 279–290). Springer-Verlag.
16. Okada, A., & Imaizumi, T. (2000). Two-mode three-way asymmetric multidimensional scaling with constraints on asymmetry. In R. Decker & W. Gaul (Eds.), *Classification and information processing at the turn of the millennium* (pp. 52–59). Springer-Verlag.
17. Okada, A., & Imaizumi, T. (2002). A generalization of two-mode three-way asymmetric multidimensional scaling. In W. Gaul & G. Ritter (Eds.), *Classification, automation, and new media* (pp. 113–122). Springer-Verlag.
18. Okada, A., & Imaizumi, T. (2003). Two-mode three-way nonmetric multidimensional scaling with different directions of asymmetry for different sources. In H. Yanai, A. Okada, K. Shigemasu, Y. Kano, & J. J. Meulman (Eds.), *New developments in psychometrics* (pp. 495–502). Springer-Verlag.
19. Okada, A., & Imaizumi, T. (2005). External analysis of two-mode three-way asymmetric multidimensional scaling. In C. Weihs & W. Gaul (Eds.), *Classification: The ubiquitous challenge* (pp. 288–295). Springer-Verlag.
20. Okada, A., & Imaizumi, T. (2005). Joint space model for multidimensional scaling of two-mode three-way asymmetric proximities. In D. Baier & K. D. Wernecke (Eds.), *Innovations in classification, data science, and information systems* (pp. 371–378). Springer-Verlag.
21. Kruskal, J. B. (1964). Nonmetric multidimensional scaling: A numerical method. *Psychometrika, 29*, 115–129.
22. Kruskal, J. B. (1964). Multidimensional scaling by optimizing goodness of fit to a nonmetric hypothesis. *Psychometrika, 29*, 1–29.
23. Kruskal, J. B., & Carroll, J. D. (1969). Geometrical models and badness-of-fit functions. In P. R. Krishnaiah (Ed.), *Multivariate analysis* (Vol. 2, pp. 639–671). Academic Press.
24. Busing, F. M. T. A., Groenen, P. J. F., & Heiser, W. J. (2005). Avoiding degeneracy in multidimensional unfolding by penalizing on the coefficient of variation. *Psychometrika, 70*, 71–98.
25. Coombs, C. H. (1950). Psychological scaling without a unit of measurement. *Psychological Review, 57*, 148–158.
26. Coombs, C. H. (1964). *A theory of data*. Wiley.
27. Heiser, W. J., & Busing, F. M. T. A. (2004). Multidimensional scaling and unfolding of symmetric and asymmetric proximity relations. In D. Kaplan (Ed.), *The Sage handbook of quantitative methodology for the social sciences* (pp. 25–48). Sage Publications.
28. Harshman, R. A. (1978). Models for analysis of asymmetrical relationships among N objects or stimuli. *Paper presented at the First Joint Meeting of the Psychometric Society and the Society for Mathematical Psychology*. McMaster University.
29. Harshman, R. A. (1981). *DEDICOM Multidimensional analysis of skew-symmetric data*. Technical memorandum, Bell Laboratories.
30. Wold, H. (1966). Estimation of principal component and related models by iterative least squares. In P. K. Krishnaiah (Ed.), *Multivariate analysis* (pp. 391–420). Academic Press.
31. Harshman, R. A., Green, P. E., Wind, Y., & Lundy, M. E. (1982). A model for the analysis of asymmetric data in marketing research. *Marketing Science, 1*, 204–242.
32. Nakayama, A. (2005). A multidimensional scaling model for three-way data analysis. *Behaviormetrika, 32*, 95–110.
33. Hayashi, C. (1972). Two-dimensional quantification based on the measure of dissimilarity among three elements. *Annals of the Institute of Mathematical Statistics, 24*, 251–257.
34. Hayashi, C. (1989). Multiway data matrix and method of quantification of qualitative data as a strategy of data analysis. In R. Coppi & S. Bolasco (Eds.), *Multiway data analysis* (pp. 131–142). North-Holland.

35. Seiyama, K., Naoi, A., Sato, Y., Tsuzuki, K., & Kojima, H. (1990). Gendai Nipponn no kaisoukouzou to sono suusei [Stratification structure of contemporary Japan and the trend]. In: A. Naoi, & K. Seiyama (Eds.), *Gendai Nippon no Kaisoukouzou Vol. 1, Shakaikaisou no kouzou to katei [Social stratification in contemporary Japan, Vol. 1, Structure and process of social stratification]* (pp. 15–50). Tokyo University Press.
36. Okada, A. (1988). An analysis of intergenerational occupational mobility by asymmetric multidimensional scaling. In M. G. H. Jansen & W. H. Schuur (Eds.), *The many faces of multivariate analysis: Proceedings of the SMABS-88 conference* (Vol. 1, pp. 1–15). RION, Institute for Educational Research, University of Groningen.
37. Slater, P. (1976). Hierarchical internal migration regions of France. *IEEE Transactions on Systems, Man and Cybernetics, 6*, 321–324.
38. Agresti, A. (2013). *Categorical data analysis* (3rd ed.). Wiley.
39. Lingoes, J. C. (1972). A general survey of the Guttman-Lingoes non-metric program series. In R. N. Shepard, A. K. Romney, & S. B. Nerlove (Eds.), *Multidimensional scaling: Theory and applications in the behavioral sciences* (pp. 49–68). Seminar Press.
40. Young, F. W. (1987). *Multidimensional scaling: History, theory, and applications.* In R. M. Hamer (Ed.). Lawrence Erlbaum Associates Inc.
41. Borg, I., & Groenen, P. J. F. (2005). *Modern multidimensional scaling. Theory and applications* (2nd edn.). Springer.
42. Groenen, P. J. F., & Heiser, W. J. (1996). The tunneling method for global optimization in multidimensional scaling. *Psychometrika, 61*(3), 529–550.